MODELS OF CELLULAR

Models of Cellular Regulation

Baltazar D. Aguda

Avner Friedman

Mathematical Biosciences Institute
The Ohio State University

OXFORD
UNIVERSITY PRESS

OXFORD
UNIVERSITY PRESS

Great Clarendon Street, Oxford, OX2 6DP,
United Kingdom

Oxford University Press is a department of the University of Oxford.
It furthers the Universitys objective of excellence in research, scholarship,
and education by publishing worldwide. Oxford is a registered trade mark of
Oxford University Press in the UK and in certain other countries

First published 2008
First published in paperback 2012

Impression: 1

British Library Cataloguing in Publication Data

Data available

Library of Congress Cataloging in Publication Data

Data available

ISBN 978–0–19–857091–2 (hbk.)
ISBN 978–0–19–965750–6 (pbk.)

Printed in Great Britain by
CPI Group (UK) Ltd, Croydon, CR0 4YY

Preface

There has been a lot of excitement surrounding the science of biology in recent years. The human genome of three billion letters has been sequenced, as have the genomes of thousands of other organisms. With unprecedented resolution, the rush of *omics* technologies is allowing us to peek into the world of genes, biomolecules, and cells – and flooding us with data of immense complexity that we are just barely beginning to understand. A huge gap separates our knowledge of the molecular components of a cell and what is known from our observations of its physiology – how these cellular components interact and function together to enable the cell to sense and respond to its environment, to grow and divide, to differentiate, to age, or to die. We have written this book to explore what has been done to close this gap of understanding between the realms of molecules and biological processes. We put together illustrative examples from the literature of mechanisms and models of gene-regulatory networks, DNA replication, the cell-division cycle, cell death, differentiation, cell senescence, and the abnormal state of cancer cells. The mechanisms are biomolecular in detail, and the models are mathematical in nature. We consciously strived for an interdisciplinary presentation that would be of interest to both biologists and mathematicians, and perhaps every discipline in between. As a teaching textbook, our objective is to demonstrate the details of the process of formulating and analyzing quantitative models that are firmly based on molecular biology. There was no attempt to be comprehensive in our account of existing models, and we sincerely apologize to colleagues whose models were not included in the book.

The mechanisms of cellular regulation discussed here are mediated by DNA (deoxyribonucleic acid). This DNA-centric view and the availability of sequenced genomes are fuelling the present excitement in biology – perhaps because one can now advance the tantalizing hypothesis that the linear DNA sequence contains the ultimate clues for predicting cellular physiology. Examples of mechanisms that explicitly relate genome structure and DNA sequence to cellular physiology are illustrated in some chapters (on gene expression and initiation of DNA replication in a bacterium); however, the majority deals with known or putative mechanisms involving pathways and networks of biochemical interactions, mainly at the level of proteins (the so-called workhorses of the cell). The quantitative analysis of these complex networks poses significant challenges. We expect that new mathematics will be developed to sort through the complexity, and to link the many spatiotemporal scales that these networks operate in. Although no new mathematics is developed in this book, we hope that the detailed networks presented here will make significant contributions to the inspiration of mathematical innovations. Another important goal is to show biologists with non-mathematical backgrounds how the dynamics of these networks are modelled, and, more importantly, to convince them that these quantitative and computational treatments are critical for progress. The collaboration between biologists and mathematical modellers is crucial in furthering our understanding of complex biological networks.

There are currently hundreds of molecular interaction and pathways databases that proliferate on the internet. In principle, these bioinformatics resources should be tapped for building or extracting models; but the sheer complexity of these datasets and the lack of automatic model-extraction algorithms are preventing modellers from using them. Although an overview of these databases is provided, almost all of the models in this book are based on current biological hypotheses on what the central molecular mechanisms of the cellular processes are.

One of us was trained as a physical-theoretical chemist, and the other as a pure mathematician. Individually, each has undergone many years of re-education and re-focusing of his research towards biology. We hope that this work will help in bringing together biologists, mathematicians, physical scientists and other non-biologists who seriously want to gain an understanding of the inner workings of life.

It is our pleasure to thank Shoumita Dasgupta for reading and generously commenting on some of the biology sections (but let it be known that lapses in biology are certainly all ours). We gratefully acknowledge the support provided by the Mathematical Biosciences Institute that is funded by the National Science Foundation USA under agreement no. 0112050.

B. D. Aguda & A. Friedman
Columbus, Ohio, USA
12 November 2007

Contents

1
General introduction

1.1 Goals

The study of life involves a bewildering variety of organisms, some extinct, while those living are in constant evolutionary flux. Amazingly, the vast spectrum of species is replaced by uniformity in composition at the molecular level. All known life forms on earth use DNA (deoxyribonucleic acid) as the carrier of information to create and sustain life – how to reproduce, how to generate energy, how to use nutrients in the environment, and how to synthesize biomolecules when needed. In recent years, high-throughput data-acquisition technologies have enabled scientists to identify and study in unprecedented detail the parts of this DNA-mediated chemical machinery – the genes, proteins, metabolites, and many other molecules.

A biological cell is a dynamic system, composed of parts that interact in ways that generate the 'living' state. Physicochemical interactions do not occur in isolation but in concert, creating pathways and networks seemingly intractable in their complexity but are somehow orchestrated to give rise to a functional unity that characterizes a living system. This book provides an account of these networks of interactions and the cellular processes that they regulate: cell growth and division, death, differentiation, and aging. The general aim is to illustrate how mathematical models of these processes can be developed and analyzed. These networks are large and require proper modelling frameworks to cope with their complexity. Such frameworks are expected to consider empirical observations and biological hypotheses that may permit network simplification. For example, living systems possess modular architecture, both in space and in terms of biological function. Modularization in space is exemplified by a cell delineated from its environment by a permeable membrane. Modularization according to biological function is another way of stating the hypothesis that – in the midst of these large, highly connected intracellular networks – only certain subnetworks are essential in driving particular cellular processes. It is the modelling of these cellular processes in terms of these subnetworks that is the subject of this book.

The cellular processes discussed here – although primarily occurring at the single-cell level – are the key determinants of cell phenotype, and therefore the physiology of the organism at the tissue level and beyond. In the next section of this general introductory chapter, some biological terms are explained and an overview of the topics covered in the chapters is given. The third section provides a general discussion of mathematical and computational modelling of biological systems. In the last section, remarks on the organization of the chapters and recommendations on how to use this book for learning, teaching and research purposes are given.

1.2 Intracellular processes, cell states and cell fate: overview of the chapters

Biology textbooks teach that the cell is the unit of life; anything less does not possess the attribute of being 'alive'. Observation of microscopic unicellular organisms – e.g. bacteria, yeast, algae – demonstrates how one cell behaves as a free-living system: it is one that grows, replicates, and responds to its environment with unmistakable autonomy and purpose. Tissues of higher animals and plants are also made up of cellular units, each with a genetic material (a set of chromosomes) surrounded by a membrane. It is this genetic material that contains the information for the replication and perpetuation of the species, and it is the localization or concentration of materials within the cell membrane that makes it possible for the operation of a 'chemical factory' that sustains life – synthesizing and processing proteins, other biomolecules, and metabolites according to the instructions encoded in the genes. Details of this picture are provided in Chapter 2 where essentials of cellular and molecular biology are summarized. This picture may be loosely called a 'genes-chemical factory' model that can begin to explain why, for instance, a muscle cell looks different from a skin cell or a nerve cell. According to this 'model', all these cells are basically the same in architecture but they look different only because of differences in the relative proportions of the proteins they make.

During the development of a multicellular organism, the *fate* of cells – that is, to what transient states or terminally differentiated states they go – depends in some complex and incompletely understood way on cell–cell and cell–environment interactions. The maturation of an organism involves multiple rounds of cell growth, division, and cell differentiation in various stages of development. Certain cells are destined to die and be eliminated in the progressive sculpting of the adult body. And in the dynamic maintenance of tissues and organs, certain cells are in continuous flux of proliferation and death – like cells of the skin and the lining of the gut. As many types of cells as there are in the adult body, there will be at least as many cell-fate decisions made. This book does not attempt to follow all these decisions (in fact, there is only one chapter that explicitly discusses cell differentiation); instead, the focus is on models of key cellular processes that impact on cell-fate decisions – gene expression, DNA replication, the cell-division cycle, cell death, cell differentiation, and cell aging (senescence). There is no attempt to be comprehensive about processes of cell-fate determination. The choice of topics is, to a large extent, dictated by the availability of published mathematical models. However, non-mathematical models – or biological hypotheses – are also discussed to anticipate biological settings for future computational modelling activities.

A prevailing biological hypothesis is that cellular 'decisions' ultimately originate from the changing states of the chromosomal DNA. Thus, cell division requires DNA replication, cell differentiation requires transcription of the DNA at select sites, and cell death is triggered when DNA damage cannot be repaired. Chapter 4 emphasizes the connectivity of gene-regulatory networks – from DNA to RNA to proteins to metabolites and back – using well-known genomic, proteomic, and metabolic information on the bacterium *Escherichia coli*. The control of the initiation of DNA replication is also well elucidated in this bacterium, and a kinetic model of this key step in cell division is discussed in Chapter 5. The importance of modelling the cell-division cycle – and also because of major recent breakthroughs in its molecular understanding – is reflected in the two chapters that follow: Chapter 6 provides a summary of the molecular machinery of the so-called 'cell-cycle engine' of eukaryotes and some recent dynamical models; Chapter 7 discusses the more complex mammalian cell cycle and its control using the mechanism of checkpoints.

Programmed cell death, also called apoptosis, is discussed in Chapter 8. Some cells are 'programmed' to die in the development of an organism or when insults on the DNA are beyond repair. As a multicellular organism grows, cells begin to acquire specific phenotypes – that is, how they look and what their functions are. Models of cell differentiation are discussed in Chapter 9. Chapter 10 deals with cell aging (senescence) and maintenance. Although there may be other mechanisms involved, the idea that there is a 'counting mechanism' for monitoring the number of times a cell divides is an intriguing one; and models have been suggested for this process. Chapter 11 deals with abnormal cell-fate regulation that leads to cancer; this last chapter illustrates tumor modelling at different scales – from intracellular pathways to cell–cell interactions in a population.

1.3   On mathematical modelling of biological phenomena

Insofar as possible, the models considered in this book are corroborated by experimental observations. The focus is on models of dynamical biological phenomena regulated by networks of molecular interactions. Model definitions range from qualitative to quantitative, or from the conceptual to the mathematical. Biologists formulate their hypotheses ('models') in intuitive and conceptual ways, often through the use of comparisons of systems observed in nature. With the aid of chemistry and physics, biological concepts and models can be couched in molecular and mechanistic terms. Just as mathematics was employed by physics to describe physical phenomena, increasingly detailed understanding of the molecular machinery of the cell is allowing the development of mechanistic and kinetic models of cellular phenomena.

A model is meant to be a replica of the system. Where details are absent – be it due to lack of instruments for direct observation or lack of ideas to explain observations – assumptions, hypotheses or theories are formulated. A scientific model involves a self-consistent set of assumptions to reproduce or understand the behavior of a system and, importantly, to offer predictions for testing the model's validity. A clear definition of the 'system' is the required first step in modelling. For example, the solar system – the sun and the eight planets – is indeed a very complex system if one includes details such as the shape and composition of the planets, but if the aim of modelling is merely to plot the trajectories of these planets around the sun, then it is sufficient to model the planets as point masses and use Newton's universal law of gravitation to calculate the planets' trajectories. It is conceivable, however, that modeling certain complex systems – such as a living cell – do not allow further simplification or abstraction below a certain level of complexity (so-called 'irreducible complexity'). Abstractions made in a model assume that certain details of the system can be 'hidden' or ignored because they are not essential in the description of the phenomenon. How such abstractions are made still requires systematic study. How can one be sure that a low-level detail is not an essential factor in the description of a higher-level system behavior? As an example where low-level property is essential for explaining higher-level behavior, one can cite the example of the anomalous heat capacity of hydrogen gas – the heat capacity being a macroscopic or system-level property – which, it turns out, can be explained by the orientation of the nuclear spins (a microscopic or low-level property) of the individual gas molecules! A similar problem arises in tumor modelling (Chapter 11) where a mutation in certain genes is eventually manifested in the behavior of cell populations in the tumor tissue. This book is about models of biological cells that are notoriously complex if one considers existing genetic and biochemical data. The premise adopted in this book is that these complex molecular networks can be modularized according to their associations with cellular processes.

In the definition of a system to be modelled, the abstraction mentioned above requires careful identification of state variables. In the example of the solar system, the state variables are the space coordinates and the velocities of the planets and the sun. Newton's laws of motion are sufficient to describe the system fully because the solutions of the dynamical equations provide the values of the state variables at any future time, given the present state of the variables. In other words, if the objective of the model is to plot planet trajectories, Newton's theory of universal gravitation provides a sufficient description of the system. What are the current physical or chemical theories upon which models of biological processes are based? As illustrated in many of the models in this book, theories of chemical kinetics are assumed to apply (these are summarized in Chapter 3). In general, existing biological models carry the implicit assumption that the fundamental principles of chemistry and physics encompass the principles necessary to explain biological behavior. There had been some serious attempts in the past to develop theories on biological processes, including theories of non-equilibrium thermodynamics and self-organizing systems (Nicolis and Prigogine, 1977). Many inorganic systems have been studied that exhibit self-organizing behavior reminiscent of living systems (Ross *et al.*, 1988), and many of these systems have been modelled using mathematical theories of non-linear dynamical systems (Guckenheimer and Holmes, 1983). The mathematical and computational methods discussed in this book are primarily those of dynamical systems theory (see Chapter 3).

What are other essential attributes of a valid biological model? There is clear evidence from detailed genetic and biochemical studies that high degrees of redundancy in the number of genes, proteins, and molecular interaction pathways are quite common in biological networks (for example, there are at least ten different cyclin-dependent kinases that influence progression of the mammalian cell cycle – see Chapter 7). This redundancy may explain the robustness of biological pathways against perturbations. Robustness is a particularly strong requirement for a valid biological model (Kitano, 2004); this is because a living cell is in a noisy environment, and key cellular decisions cannot be at the mercy of random fluctuations. This robustness requirement on biological models translates either to robustness against perturbations of model parameters, or against perturbations of edges – that is, adding or deleting interactions – in the network.

Lastly, a mathematical model must lend itself to experimental verification. Given a set of experimental data, a modeller is faced with the difficulty of enumerating possible models that can explain the data. A proposed model must offer predictions and explicit experimental means to discriminate itself from other candidate models. This iterative process between model building and experimental testing represents the essence of scientific activity.

1.4   A brief note on the organization and use of the book

This book is addressed to students of the mathematical, physical, and biological sciences who are interested in modelling cellular regulation at the level of molecular networks. Where the mathematics could be involved (but is interesting to non-biologists who may wish to pursue the topics further), sections indicated by ■ can be omitted on first reading.

Chapters 2 to 4 form the foundations on the biology and mathematical modelling approaches used in the entire book. Although Chapter 2 is a very brief summary of essential cellular and molecular biology, it embodies the authors' perspective on what aspects of the biology are essential in modelling. Chapter 3 is a summary of key mathematical modelling tools and guides the reader to more detailed modelling resources; more importantly, this chapter explains how models are created and set up for analysis.

The remaining chapters can be read independently, although it is recommended that Chapters 6 and 7 be read in sequence. The arrangement of the chapters, however, was conceived by the authors to develop a story about the regulation of cellular physiology – gene expression and cell growth, gene replication and cell division, death, differentiation, aging, and what happens when these processes are compromised in cancer. A glossary of terms and phrases is included at the end of the book.

References

Guckenheimer, J. and Holmes, P. (1983) *Nonlinear oscillations, dynamical systems, and bifurcation of vector fields.* Springer Verlag, New York.

Kitano, H. (2004) 'Biological Robustness', *Nature Reviews Genetics* **5**, 826–837.

Nicolis, G. and Prigogine, I. (1977) *Self-organization in nonequilibrium systems.* Wiley, New York.

Ross, J., Muller, S. C. and Vidal, C. (1988) 'Chemical Waves', *Science* **240**, 460–465.

2
From molecules to a living cell

One of the striking features of life on earth is the universality (as far as we know) of the chemistry of the basic building blocks of cells; this is especially true in the case of the carrier of genetic information, the DNA. This universality suggests that it is in the intrinsic physicochemical properties of these biomolecules where one can find the origins of spatiotemporal organization and functions characteristic of living systems. At the level of molecular interactions, fundamental laws of physics and chemistry apply. However, the emergence of the 'living state' is expected to be associated with ensembles of molecular processes organized spatially in organelles and other cellular compartments, as well as temporally in their dynamics *far from equilibrium*. To help understand these levels of organization, the basic anatomy of cells, the properties of these biomolecules and their interactions are summarized in this chapter. Of central importance is the molecular machinery for expressing genes to proteins; this is a complex but well-orchestrated machinery involving webs of gene-interaction networks, signalling and metabolic pathways. Information on these networks is increasingly and conveniently made available in public internet databases. A brief survey is given at the end of this chapter of the major databases containing genomic, proteomic, metabolomic, and interactomic information. The challenge to scientists for decades to come is to integrate and analyze these data to understand the fundamental processes of life.

2.1   Cell compartments and organelles

A diagram of the basic architecture of *eukaryotic* cells is shown in Fig. 2.1. Every eukaryotic cell has a membrane-bound *nucleus* containing its chromosomes. In contrast, a *prokaryotic* cell lacks a nucleus; instead, the chromosome assembly is referred to as a *nucleoid*. A description of the compartments and major organelles in a representative eukaryotic cell is given in this section.

A bilayer phospholipid membrane, called the *plasma membrane*, delineates the cell from its environment. This membrane allows the selective entry of raw materials for the synthesis of larger biomolecules, the transmission of extracellular signals (e.g. from extracellular ligands docking on membrane-receptor proteins), retains or concentrates substances needed by the cell, and the efflux of waste products. Each phospholipid molecule has a hydrophobic (or 'water-hating') end and a hydrophilic (or 'water-loving') end. When these molecules are dispersed in water, they aggregate spontaneously to form a bilayer membrane, both surfaces of the membrane being lined

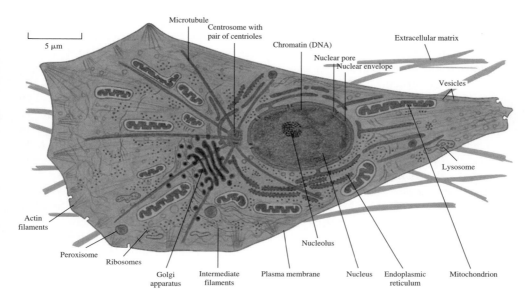

Fig. 2.1 The major compartments and organelles of a typical eukaryotic cell. The plasma membrane, chromosomes (condensed chromatin), ribosomes, nucleolus, mitochondria, centrosome and the cytoskeleton (microtubules and filaments) are described in the text. The Golgi apparatus is referred to as the 'post office' of the cell: it 'packages' and 'labels' the different macromolecules synthesized in the cell, and then sends these out to different places in the cell. Lysosomes are organelles containing digestive enzymes, which is why they are also called 'suicide sacs' because spillage of their contents causes cell death. Reproduced with permission from the book of Alberts *et al.* (2002). © 2002 by Bruce Alberts, Alexander Johnson, Julian Lewis, Martin Raff, Keith Roberts, and Peter Walter. (See Plate 1)

by the hydrophilic ends of the lipid molecules, while the hydrophobic ends are tucked in between the surfaces. This is an example of a common observation that many types of biomolecules synthesized by cells possess the ability to self-assemble into structures with specific cellular functions (other examples will be given below).

Proteins that span the plasma membrane, called *transmembrane proteins*, are involved in cell–environment and cell–cell communications. Examples of these proteins are *ion-channel* proteins (e.g. sodium and potassium ion channels involved in regulating the electric potential difference across the plasma membrane) and *membrane-receptor* proteins, whose conformational changes (brought about, for example, by binding with extracellular ligands) usually initiate cascades of biochemical processes that get transduced to the nuclear DNA causing changes in gene expression. Certain membrane proteins are involved in cell–cell recognition that is crucial in the operation of the immune system.

The material between the plasma membrane and the nucleus is called the *cytoplasm*. Encased by the nuclear membrane are the chromosomes that contain the *genome* (set of genes) of the organism. Humans (*Homo sapiens*) have 46 chromosomes in their somatic cells. Human sperm and egg cells have 23 chromosomes each.

Although the code for producing proteins is in the chromosomes, proteins are synthesized outside the nucleus in sites that look like granules under the microscope. These sites of protein synthesis are the *ribosomes* (see Fig. 2.1 and Fig. 2.2). As shown in Fig. 2.1, ribosomes are either attached to a network of membranes (called the *endoplasmic reticulum*) or are free in the cytoplasm. A bacterium such as *E. coli* cell has $\sim 10^4$ ribosomes and a human cell has $\sim 10^8$ ribosomes. The assembly of ribosomes originates from a nuclear compartment called the *nucleolus* (see Fig. 2.1).

Besides proteins, many other types of molecules are produced in the cell through enzyme-catalyzed *metabolic* reactions. The organelles called *mitochondria* (Fig. 2.1) are the cell's power plants because most of the energetic molecules – called ATP (*adenosine tri*phosphate) – are generated in these organelles. Energy is released when a phosphate bond is broken during the transformation of ATP into ADP (*adenosine di*phosphate); this energy is used to drive many metabolic reactions. A typical eukaryotic cell contains ~ 2000 mitochondria. (Interestingly, mitochondria contain DNA, which suggests – according to the endosymbiotic theory – that these organelles were once free-living prokaryotes.)

As depicted in Fig. 2.1, the shape of the cell is maintained by the *cytoskeleton* that is a network of *microtubules* and *filaments*. These cytoskeletal elements are self-assembled from smaller protein subunits. Rapid disassembly and assembly of these subunits can occur in response to external signals (this happens, for example, when a cell migrates). Of major importance to cell division is the organelle called *centrosome* that is composed of a pair of barrel-shaped microtubules called *centrioles* (Fig. 2.1). Immediately after the chromosomes are duplicated, the centrosome is also duplicated; the two centrosomes are eventually found in opposite poles prior to cell division. The spindle fibers (microtubules) emanating from these two centrosomes carry out the delicate task of segregating the chromosomes equally between daughter cells.

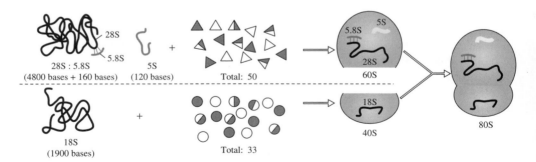

Fig. 2.2 Ribosomes of mammalian cells. Shown are schematic pictures of the components of the large (60S) and small (40S) subunits of the ribosome (80S). The strands represent ribosomal RNAs, and the triangles are the 50 proteins of the large subunit and the 33 proteins of the small subunit. Figure reproduced with permission from Lodish *et al. Molecular cell biology.* © 2000 by W. H. Freeman and Company.

The components and structures of cell organelles and other large protein complexes have been elucidated. For example, mammalian ribosomes are large complexes of 83 proteins and 4 ribonucleic acids (see Fig. 2.2). Other important examples are the components and the mechanisms of action of various polymerase enzymes in the replication of chromosomes (DNA polymerases) and in decoding genes (RNA polymerases). Many of these macromolecular complexes are being viewed as molecular machines.

To reiterate, a wide variety of the biomolecules synthesized in cells self-assemble spontaneously. The phospholipid molecules of the plasma membrane – products of cell metabolism – form bilayers spontaneously in aqueous solutions. In the construction of the cytoskeleton, tubulin proteins polymerize to form microtubules, actin to microfilaments, and myosin to thick filaments. Recent studies even suggest that the whole eukaryotic nucleus is a self-assembling organelle.

2.2   The molecular machinery of gene expression

All known living things on earth use DNA (*deoxyribonucleic acid*) as the genetic material (except for some viruses that use *ribonucleic acid* or RNA for short). The publication of the structure of DNA by James Watson and Francis Crick in 1953 revolutionized biology. The structure of DNA provides a clear molecular basis for the inheritance of genes from one generation to the next, as described in more detail below.

In each eukaryotic chromosome, DNA exists as two strands paired to form a double helix (Fig. 2.3). Each strand has a sugar–phosphate backbone, and attached to the sugars are four nitrogenous bases, namely, adenine (A), thymine (T), cytosine (C), and guanine (G). The double helix is formed from the Watson–Crick pairing between these bases: A paired to T, and C paired to G. As shown schematically in Fig. 2.3, the specificity of these pairings is due to the number of hydrogen bonds between the bases. Because these hydrogen bonds are weak – unlike the much stronger covalent bonds in molecules – they allow the 'unzipping' of the double helix during DNA replication. Note that the T–A pair has two hydrogen bonds while the G–C pair has three, suggesting that the double helix is easier to unzip where there are more T–A pairs than G–C pairs. It is these Watson–Crick base pairings that elegantly explain the molecular basis of gene inheritance.

For DNA replication to start, the duplex has to 'unzip' to expose single-stranded DNA segments where synthesis of new DNA strands occur according to the Watson–Crick base pairing. This is a highly regulated affair involving dozens of enzymes, including DNA polymerases.

Genes correspond to stretches of sequences of the letters A, T, C, G on the DNA (DNA segments comprising a gene are not necessarily contiguous). *Gene expression* refers to the synthesis of the protein according to the DNA sequence of the gene (also called protein-coding sequence). The gene-expression machinery requires that the DNA sequence is first transcribed to an RNA sequence. RNA molecules also have the A, C, and G bases, but uracil (U) is used instead of T. RNA molecules do not stably form double helices like DNA. However, the pairings of C–G and A–U are observed. The gene-expression machinery is summarized in Fig. 2.4.

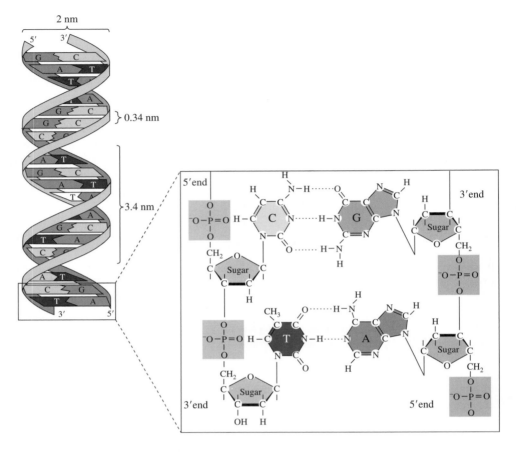

Fig. 2.3 Two DNA molecules form the Watson–Crick double helix where the sugar–phosphate backbones are on the outside and the bases are inside, paired by hydrogen bonds as shown on the right of the figure (A with T, and C with G). The 5′ and 3′ designations of the ends of a DNA strand are based on the numbering of the C atoms on the deoxyribose (sugar). Figure reproduced with permission from: G. M. Cooper and R. Hausman, (2007) *The cell: a molecular approach.* 4th edn. © ASM Press and Sinauer Assoc., Inc.

As depicted in Fig. 2.4, the DNA double helix is unzipped where particular genes are located so that the enzyme called RNA polymerase can transcribe the DNA sequence into RNA. This primary RNA contains sequences called *exons* and *introns*; the latter do not code for proteins and are removed. The remaining exons are then stitched together through a process called *RNA splicing* to form a continuous molecule of mature *messenger RNA* (mRNA). This mRNA relocates from the nucleus to the cytoplasm where it is *translated* in ribosomes. Thus, gene expression is defined as the combination of transcription and translation to the protein product.

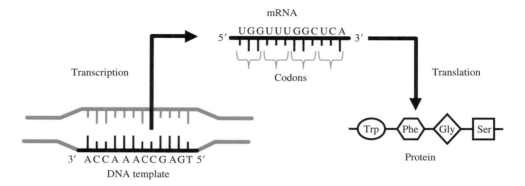

Fig. 2.4 Gene expression is carried out in two steps: *transcription* of DNA to RNA, followed by *translation* of the messenger RNA (mRNA) to protein. The correspondence between a *codon* (a triplet of bases) and the translated amino acid is given by the genetic code (Table 2.1).

A key question is the correspondence between the mRNA sequence and the amino-acid sequence of the protein product. One of the triumphs of molecular genetics is the discovery of the universal *genetic code* shown in Table 2.1. The genetic code gives the correspondence between *codons* (three-nucleotide sequences) on the mRNA and the 20 amino acids found in almost all naturally occurring proteins. There is a total of 4^3 or 64 possible codons, all listed in Table 2.1. The code also specifies codons that signal termination and initiation of translation. The code is degenerate in the sense that more than one codon can specify a single amino acid (but not vice versa). As depicted in Fig. 2.5, small RNAs (composed of 73 to 93 nucleotides) called transfer RNAs (tRNAs) act as adaptor molecules that read the mRNA codons. Each tRNA has a sequence of three nucleotides called an *anticodon* that matches the mRNA codon by Watson–Crick complementarity. The ribosome moves along the mRNA, and the charged tRNAs (i.e. those carrying their specific amino acids) enter in the order specified by the mRNA codons (see Fig. 2.5). The contiguous amino acids are then enzymatically joined to form polypeptides (proteins).

One can conclude that the amino-acid sequences of all cellular proteins are encoded in the DNA. Changes in certain DNA sequences can have drastic consequences on the shape and function of translated proteins. For example, a particular mutation in the hemoglobin gene (namely, a specific GAG sequence in the DNA is changed to GTG) leads to the disease called sickle-cell anemia; here, the corresponding single amino-acid change causes a drastic change in the shape of hemoglobin that compromises the protein's function as carrier of oxygen in red blood cells. The shape of proteins largely determines their biological functions, giving a rationale to many observations that, in the course of evolution, the three-dimensional structures of proteins are better conserved than their one-dimensional amino-acid sequences. Although many advances have been made recently, the problem of predicting three-dimensional structures of proteins from their one-dimensional amino-acid sequence is still not solved.

Table 2.1 The genetic code: from RNA codons to amino acids. A 'stop' codon signifies termination of translation. AUG (Met) is the usual initiator codon, but CUG and GUG are also used as initiator codons in rare instances. The 3-letter symbols in this table are for the following amino acids: L-Alanine (Ala), L-Arginine (Arg), L-Asparagine (Asn), L-Aspartic acid (Asp), L-Cysteine (Cys), L-Glutamic acid (Glu), L-Glutamine (Gln), Glycine (Gly), L-Histidine (His), L-Isoleucine (Ile), L-Leucine (Leu), L-Lysine (Lys), L-Methionine (Met), L-Phenylalanine (Phe), L-Proline (Pro), L-Serine (Ser), L-Threonine (Thr), L-Tryptophan (Trp), L-Tyrosine (Tyr), L-Valine (Val).

First position (5' end)	Second position: U	C	A	G	Third position (3' end)
U	Phe	Ser	Tyr	Cys	U
	Phe	Ser	Tyr	Cys	C
	Leu	Ser	stop	stop	A
	Leu	Ser	stop	Trp	G
C	Leu	Pro	His	Arg	U
	Leu	Pro	His	Arg	C
	Leu	Pro	Gln	Arg	A
	Leu (Met)	Pro	Gln	Arg	G
A	Ile	Thr	Asn	Ser	U
	Ile	Thr	Asn	Ser	C
	Ile	Thr	Lys	Arg	A
	Met (start)	Thr	Lys	Arg	G
G	Val	Ala	Asp	Gly	U
	Val	Ala	Asp	Gly	C
	Val	Ala	Glu	Gly	A
	Val (Met)	Ala	Glu	Gly	G

2.3 Molecular pathways and networks

Although many of the so-called *housekeeping genes* are constitutively expressed for cell maintenance, there are also many other genes whose expressions respond or adapt to conditions of the cell environment. As a specific example, the bacterium *E. coli* can synthesize tryptophan (Trp) if the level of this amino acid in the extracellular medium is low; otherwise the bacterium shuts off its endogenous Trp-synthesizing machinery. The network of molecular interactions regulating Trp synthesis, from the transcription and translation of genes to the metabolic pathway that generates the amino acid, will be analyzed in Chapter 4. The Trp network is a good example of how the expression of genes can be affected by their products – thus forming feedback loops in the network.

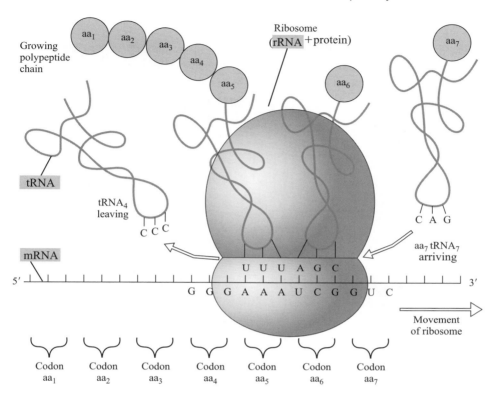

Fig. 2.5 A cartoon of how the ribosome moves along the mRNA to translate the codons to amino acids – in collaboration with tRNAs that are charged with corresponding amino acids (*circles* labelled *aa*s in the diagram). Figure reproduced with permission from Lodish *et al.* *Molecular cell biology.* © 2000 by W. H. Freeman and Company.

The metabolic steps in the synthesis of the 20 amino acids in the universal genetic code, as well as other essential biomolecules – nucleotides, lipids, carbohydrates and many others – are coupled in a complex web of metabolic reactions. The steps in the metabolism of these biomolecules require enzymes (proteins) to occur, and therefore one can claim that the set of biochemical reactions in a cell is orchestrated by the information contained in its genome. A glimpse of the complexity of metabolic pathways is shown in Fig. 2.6.

In addition to metabolic networks, many other cellular networks involve the regulation of the activities of enzymes and other proteins. Enzymes are found in both inactive and active states, and the switching between these states involve regulatory networks whose complexity may reflect the importance of the enzyme function. These post-translational protein networks add another layer in the complexity of cellular networks. Figure 2.7 is a broad summary of these networks as they relate to the 'DNA-to-RNA-to-protein' flow of information; the general network shown in the

METABOLIC PATHWAYS

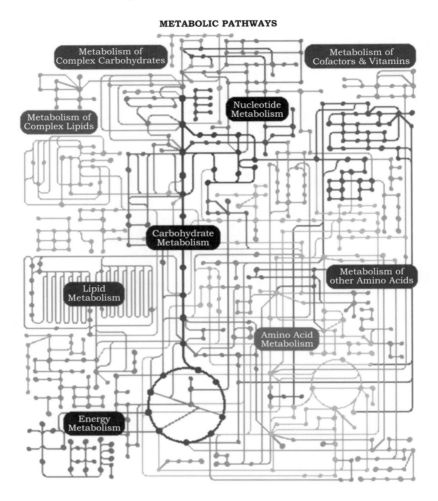

Fig. 2.6 Metabolic pathways from the online database KEGG (Kyoto Encyclopedia of Genes and Genomes, see Table 2.2 for its internet address). Each dot in the above 'wiring' diagram represents a metabolite (usually a small organic molecule). The edge between dots represents a chemical reaction that is catalyzed by an enzyme (which, in turn, is usually synthesized by a cell's gene-expression machinery). Figure reproduced with permission from KEGG (Courtesy of Prof. M. Kanehisa).

figure is referred to in this book as *gene-regulatory networks* (GRNs). As indicated by the many feedback loops in this diagram, the information flow is not strictly linear; for example, reverse transcription from RNA to DNA is accomplished by retroviruses. Feedback loops may occur at every step during gene expression where

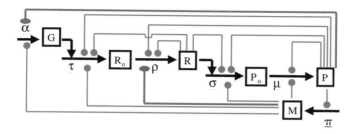

Fig. 2.7 A broad summary of gene-regulatory networks. The *arrow* labelled π represents metabolic networks requiring proteins (P) to catalyze reactions that produce metabolites (M); in general, metabolites are needed in every step of the gene-expression machinery. The *arrow* labelled α represents the replication of the genomic DNA – a process that needs metabolites (nucleotides), proteins (polymerases), and RNA (edge from R is not shown in figure). Transcriptional units (G) in the genome are transcribed in step τ to primary RNA transcripts (R_o) that are processed in step ρ to form mature transcripts (R). Proteins, such as transcription factors, can directly influence the transcription step τ. The translation of mRNAs to proteins (P_o) in step σ requires the co-operation of many proteins, tRNAs, and ribosomal RNAs. Step μ represents post-translational modifications of proteins that render them functional. The edges that end in dots (regulating the steps in the network) represent either activatory or inhibitory influence.

products can influence the rates of information flow, as well as which information is to be transmitted.

The existence of the many feedback loops depicted in Fig. 2.7 presents a formidable challenge in the analysis of GRNs. Many chapters in this book deal with models that implicitly assume a modularization of these large cellular networks – that is, focusing only on subnetworks that are assumed to explain particular cellular phenomena or functions. This reductionist approach is open to question in light of the highly connected property of cellular networks.

2.4   The *omics* revolution

A draft of the human genome sequence was first published in 2001 and a more complete version was generated in 2003 (online educational resources can be found at http://genome.gov/HGP/). This sequence is that of the approximately 3 billion 'letters' (bases) consisting of A, T, C, and G in human DNA. Current estimates of the number of human genes range from 20 000 to 30 000. An internet portal for databases on genomic sequences (DNA and RNA) is the website of the *International Nucleotide Sequence Database Collaboration* (http://www.insdc.org/) linking websites in the USA, Japan and Europe. To date, over 100 gigabases of DNA and RNA sequences have been deposited in public internet databases. These sequences represent individual genes and partial or complete genomes of more than 165 000 organisms.

Table 2.2 A few major pathway and modelling resources on the internet. For a more comprehensive list, go to the *Pathguide* website address given in this table.

Pathway & modelling resources	URL
General web portal	
Pathguide	http://pathguide.org
Ontologies	
Gene Ontology	http://www.geneontology.org
BioPAX	http://www.biopax.org
Pathway maps	
KEGG	http://www.genome.jp/kegg/
Reactome	http://www.reactome.org
GenMAPP	http://genmapp.org
Biocarta	http://biocarta.com
Pathway Interaction Database	http://pid.nci.nih.gov/
Model repositories	
Biomodels	http://www.ebi.ac.uk/biomodels/
CellML	http://www.cellml.org

Transcriptomics – the measurement of RNA transcript levels on a large scale – is made possible by high-throughput microarray technologies. Proteomics technologies currently being developed aim for the identification and measurement of all proteins in a cell. Similarly, metabolomic technologies aim at analyzing metabolites. These so-called *omics* technologies are providing a comprehensive 'parts list' of the cell. However, in order to understand how the cell works, it is necessary to determine and understand the interactions among these parts. The preceding sections have provided glimpses into the complexity of these networks of interactions. Table 2.2 gives a short list of the major internet resources on pathways databases. The website called *Pathguide* is a good internet portal to more than 200 of these databases. In addition to the literature (of which *Pubmed* is an important electronic resource, http://pubmed.gov), these pathways databases are important resources for modeling cellular processes. *Gene Ontology* and *BioPAX* represent efforts in the bioinformatics community to standardize the annotation of genes and of pathways, respectively. The websites listed under *Pathway Maps* in Table 2.2 are good sources of diagrams of many cellular pathways. The websites *Biomodels* and *CellML* are repositories of published mathematical models of a diverse range of cellular processes.

References & further readings

Alberts, B., Johnson, A., Lewis, J., Raff, M., Roberts, K., and Walter, P. (2002) *Molecular biology of the cell.* 4th edn. Garland Science, New York.

Cooper, G. M. and Hausman, R. E. (2006) *The cell: a molecular approach.* 4th edn. Sinauer Associates, Inc., Sunderland, MA, USA.

Harold, F. M. (2001) *The way of the cell: molecules, organisms and the order of life.* Oxford University Press, New York, NY.

Judson, H. F. (1996) *The eighth day of creation: makers of the revolution in biology.* Cold Spring Harbor Laboratory Press.

Lodish, H., Berk, A., Zipursky, S. L., Matsudaira, P., Baltimore, D., and Darnell, J. E. (1999) *Molecular cell biology.* W. H. Freeman & Co., New York, NY.

3
Mathematical and computational modelling tools

Most of the mathematical equations in this book are descriptions of the dynamics of biochemical reactions and associated physical processes. A brief review of chemical kinetics is therefore provided in this chapter to illustrate the formulation of model equations for a given reaction mechanism. For spatially uniform systems, these model equations are usually ordinary differential equations; but coupling of chemical reactions to physical processes such as diffusion requires the formulation of partial differential equations to describe the spatiotemporal evolution of the system. Mathematical analysis of the dynamical models involves basic concepts from ordinary and partial differential equations (such as bifurcation and stability) that are reviewed in this chapter. Computational methods, including stochastic simulations, and sources of computer software programs available free on the internet are also summarized.

3.1   Chemical kinetics

Suppose a molecule of A and two molecules of B react to form a new molecule C. Chemists depict this reaction as follows

$$A + 2B \xrightarrow{k_1} C, \tag{3.1}$$

where k_1 is called the *rate coefficient* (sometimes called *rate constant*). As shown in the chemical equation, the *stoichiometric coefficients* of A, B and C are 1, 2 and 1, respectively. The respective concentrations of these molecules are denoted by [A], [B], and [C]. The *law of mass action* states that the rate of a given reaction is proportional to the concentrations of the chemical species as written on the reactant (left) side of the chemical equation; thus, for reaction 3.1 whose reactant side can be written as $A + B + B$, the reaction rate v_1 is equal to $k_1[A][B]^2$. Thus,

$$\frac{d[C]}{dt} = v_1,$$

$$\frac{d[A]}{dt} = -v_1,$$

$$\frac{d[B]}{dt} = -2v_1.$$

One can write the corresponding reversible reaction as two one-way reactions:

$$A + 2B \xrightarrow{k_1} C, \quad C \xrightarrow{k_2} A + 2B. \tag{3.2}$$

With the rate of the reverse reaction $v_2 = k_2[C]$, the kinetic equations are now the following:

$$\frac{d[C]}{dt} = v_1 - v_2,$$

$$\frac{d[A]}{dt} = -v_1 + v_2,$$

$$\frac{d[B]}{dt} = -2v_1 + 2v_2.$$

Note that the above equations can be written in the following vector-matrix form

$$\frac{d}{dt} \begin{bmatrix} A \\ B \\ C \end{bmatrix} = \begin{bmatrix} -1 & 1 \\ -2 & 2 \\ 1 & -1 \end{bmatrix} \begin{bmatrix} v_1 \\ v_2 \end{bmatrix}. \tag{3.3}$$

The discussion above can be generalized to a system with an arbitrary number of chemical reactions. Let there be n chemical species whose concentrations are $[X_1]$, $[X_2]$, ..., $[X_n]$. Let there be r chemical reactions, with each reaction being symbolized by

$$s^R_{1j}X_1 + s^R_{2j}X_2 + \cdots + s^R_{nj}X_n \longrightarrow s^P_{1j}X_1 + s^P_{2j}X_2 + \cdots + s^P_{nj}X_n \quad (j = 1, \ldots, r), \tag{3.4}$$

where s^R_{ij}, s^P_{ij} are the stoichiometric coefficients of species i on the reactant side and product side, respectively, of reaction j. Let $\mathbf{X}$ be the concentration vector $[[X_1], \ldots, [X_n]]$, $\mathbf{v}$ the reaction velocity vector $[v_1, \ldots, v_r]$ where v_j is the rate of reaction j, and $\mathbf{S}$ the so-called *stoichiometric matrix* whose element s_{ij} is equal to $(s^P_{ij} - s^R_{ij})$. The general set of dynamical equations for chemical reactions systems can be written succinctly as

$$\dot{\mathbf{X}} = \mathbf{Sv}, \tag{3.5}$$

where $\dot{\mathbf{X}}$ means $d\mathbf{X}/dt \equiv [d[X_1]/dt, \ldots, d[X_n]/dt]$. Such a system of ODEs is called a *stoichiometric dynamical system*. One expects that the stoichiometric matrix $\mathbf{S}$ exerts a considerable constraint on the dynamics of the system. For readers interested in pursuing this topic in detail, the works of Feinberg and of Clarke are recommended (see references).

A reaction j is said to have *mass-action kinetics* if its rate v_j has the form

$$v_j = k_j \prod_{i=1}^{n} X_i^{s^R_{ij}}.$$

Because this kinetics is based on probabilities of collisions between reactant molecules, the so-called *order of the reaction* – defined as $\sum_{i=1}^{n} s_{ij}^{R}$ for the jth reaction (see eqn 3.4) – is usually equal to 1 or 2, and rarely 3. Note that there are other types of kinetics, as discussed below.

Consider the enzyme-catalyzed conversion of a substrate S into a product P:

$$S \xrightarrow{E} P, \tag{3.6}$$

where the E on top of the arrow is the enzyme catalyzing the reaction. How the enzyme interacts with the substrate to generate the product is not shown explicitly in eqn 3.6 – this reaction only gives the overall reaction, and the enzyme is regenerated after the reaction. Many one-substrate enzymatic reactions of the type in eqn 3.6 show initial rates that follow *Michaelis–Menten kinetics* of the form

$$v = \frac{V_{\max}[S]}{K_M + [S]}, \tag{3.7}$$

where [S] is the substrate concentration, $V_{\max}$ is the maximum rate of the reaction, and K_M is the *Michaelis constant*. A plot of v versus [S] is shown in Fig. 3.1 by the curve with $n = 1$.

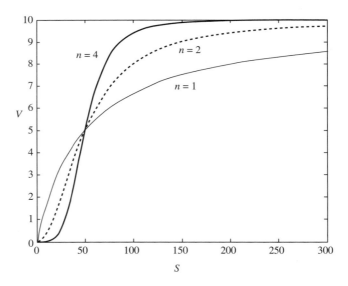

Fig. 3.1 Rate of an enzyme reaction versus substrate concentration according to the Michaelis–Menten kinetics ($n = 1$) and Hill-type kinetics ($n = 2, 4$). The three curves are generated from the equation $v = V_{\max} S^{n}/(K_M^{n} + S^{n})$ with $V_{\max} = 10$ and $K_M^{n} = 50$.

One possible mechanism that gives rise to the rate expression in eqn 3.7 is the following (see Exercise 1):

$$S + E \xrightarrow{k_1} ES$$

$$ES \xrightarrow{k_{-1}} S + E \qquad (3.8)$$

$$ES \xrightarrow{k_2} E + P.$$

Note that the total enzyme concentration, $E_{\text{tot}} = [E] + [ES]$, is a constant.

Another typical rate expression in enzyme kinetics is the *Hill function* given by

$$v = \frac{c_1 [S]^n}{c_2 + [S]^n} \qquad (n > 1). \qquad (3.9)$$

As shown in Fig. 3.1, as n increases the rate becomes more sigmoidal in shape. An example of a mechanism giving rise to eqn 3.9, for the specific case of $n = 2$, is the following:

$$S + E \xrightarrow{k_1} ES, \quad ES \xrightarrow{k_{-1}} S + E$$

$$ES \xrightarrow{k_2} E + P$$

$$\qquad (3.10)$$

$$S + ES \xrightarrow{k_3} ES_2, \quad ES_2 \xrightarrow{k_{-3}} S + ES$$

$$ES_2 \xrightarrow{k_4} ES + P.$$

Observe that the net overall reaction (i.e. involving only S and P) is identical to 3.6. The overall reaction rate is $v = d[P]/dt = v_2 + v_4 = k_2[ES] + k_4[ES_2]$. The steady-state approximations $d[ES]/dt = d[ES_2]/dt = 0$ lead to the following expression

$$v = \frac{(k_2 K_2 + k_4 [S]) E_{\text{tot}} [S]}{K_1 K_2 + K_2 [S] + [S]^2}, \qquad (3.11)$$

where $K_1 = (k_{-1} + k_2)/k_1$ and $K_2 = (k_{-3} + k_4)/k_3$. Note that steps 1 and 3 in eqn 3.10 represent sequential binding of two substrate molecules to the enzyme. *Co-operativity* is said to exist if a previously enzyme-bound substrate molecule significantly increases the rate of binding of a second substrate molecule; this is the case when $k_3 \gg k_1$. The extreme case of $k_1 \to 0$ and $k_3 \to \infty$, with $k_1 k_3$ a finite positive number, leads to $K_1 \to \infty$ and $K_2 \to 0$ with $K_1 K_2$ a finite positive constant; under these conditions, eqn 3.11 becomes

$$v = \frac{V_{\text{max}} [S]^2}{K_{\text{M}}^2 + [S]^2}, \qquad (3.12)$$

where $V_{\text{max}} = k_4 E_{\text{tot}}$ and $K_{\text{M}}^2 = K_1 K_2$.

It can be shown that if n substrate molecules bind sequentially to the enzyme, with corresponding constants K_1 to K_n, then in the limit $K_n \to 0$ and $K_1 \to \infty$

(with $K_1 K_n$ constant), the overall rate of the reaction is given by

$$v = \frac{V_{\max}[S]^n}{K_M^n + [S]^n}. \tag{3.13}$$

This equation is the general *Hill equation*.

3.2 Ordinary differential equations (ODEs)

The chemical kinetic equations discussed in the previous section assume that the system is spatially homogeneous, and that temperature and pressure are fixed. The equations are systems of nonlinear ordinary differential equations (ODEs) of the form:

$$\frac{dx_i}{dt} = f_i(x_1, \ldots, x_n), \quad i = 1, \ldots, n,$$

or, in vector form,

$$\frac{dx}{dt} = f(x), \tag{3.14}$$

where $x = (x_1, \ldots, x_n)$ and $f = (f_1, \ldots, f_n)$. In this section some useful general properties of ODEs are reviewed, including existence, uniqueness, and stability of solutions. Subsequent sections (Sections 3.3 to 3.7) develop the notions of phase space, bifurcation and stability in portraying the dynamics of these systems. These mathematical considerations are important in assessing the validity of biological models.

■ 3.2.1 Theorems on uniqueness of solutions

In standard chemical kinetic theory, the functions f_i and their derivatives are assumed to be continuous because the rate of each reaction step is assumed to be a continuous function with continuous derivatives. If the first derivative is not continuous then the uniqueness of the solution $x(t)$ of the ODE system eqn 3.14 is not guaranteed. The following theorems are well known:

Theorem 3.1 *If the functions $f_i(x)$ in eqn 3.14 and their partial derivatives $\partial f_i(x)/\partial x_j$ are continuous for $-\infty < x_k < +\infty$ ($k = 1, \ldots, n$), then for any initial condition*

$$x(0) = x_0 \tag{3.15}$$

there exists a unique solution $x(t)$ of the system of eqns 3.14 and 3.15 for some small time interval $-\delta < t < \delta$.

As an example where Theorem 3.1 does not apply, consider the following equation

$$\frac{dx}{dt} = x^\alpha \ (0 < \alpha < 1), \quad x(0) = 0.$$

The derivative of the right-hand side of the equation at $x(0)$ goes to infinity, and the system does not have a unique solution. Indeed, $x(t) \equiv 0$ is one solution and another solution is $x(t) = Ct^{\frac{1}{1-\alpha}}$ where $C^{1-\alpha} = 1 - \alpha$.

Theorem 3.2

(i) If $\mathbf{f}(\mathbf{x})$ is bounded linearly, that is

$$|\mathbf{f}(\mathbf{x})| \leq c_1 |\mathbf{x}| + c_2 \tag{3.16}$$

for some positive constants c_1 and c_2, then the solution can be uniquely extended to all $-\infty < t < \infty$.

(ii) If $\mathbf{f}(\mathbf{x})$ is not bounded linearly, but

$$\mathbf{x} \cdot \mathbf{f}(\mathbf{x}) \leq c_1 |\mathbf{x}|^2 + c_2 \tag{3.17}$$

for some positive constants c_1 and c_2, then the solution $\mathbf{x}(t)$ can uniquely be extended for $-\delta < t < \infty$.

In the last theorem, the following notations were used:

$$|\mathbf{z}| = \left(\sum_{i=1}^{n} z_i^2\right)^{1/2} \quad \text{for } \mathbf{z} = (z_1, \ldots, z_n), \text{ and } \mathbf{x} \cdot \mathbf{z} = \sum_{i=1}^{n} x_i z_i.$$

3.2.2 Vector fields, phase space, and trajectories

A set of solution $\mathbf{x}(t)$ of eqn 3.14, for all time t, emanating from the initial condition $\mathbf{x}_0$ is called a *trajectory*. The solution $\mathbf{x}(t)$ is sometimes referred to as the *state* of the system at time t. The space of all possible states $\mathbf{x} = (x_1, \ldots, x_n)$ is called the *phase space* or *state space*. The right-hand side of eqn 3.14 is called a *vector field* because it assigns to each point $\mathbf{x}$ in the phase space a direction of the flow of trajectories; in other words, a trajectory $\mathbf{x}(t)$ is simply a curve with the tangent vector at each time t given by the vector $\mathbf{f}(\mathbf{x}(t))$ (see Fig. 3.2(a)). Another name for a trajectory that exists for all $-\infty < t < +\infty$ is *orbit*. Figure 3.2(b) shows three trajectories from

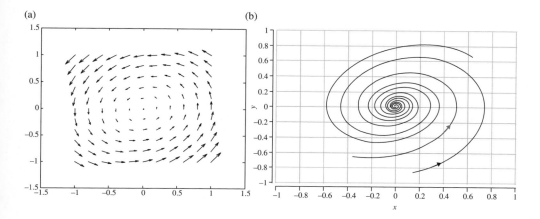

Fig. 3.2 (a) A vector field and (b) trajectories associated with the system $dx/dt = -y$, $dy/dt = x + cy$, $c = -0.3$.

different initial conditions. A trajectory, or orbit, is called *periodic* with period T if $\mathbf{x}(t+T) = \mathbf{x}(t)$ for $-\infty < t < +\infty$.

3.2.3 Stability of steady states

A point $\mathbf{x}_0$ such that $\mathbf{f}(\mathbf{x}_0) = \mathbf{0}$ is called a *steady state* (also called an *equilibrium point*, or a *fixed point*) of the ODE eqn 3.14. A steady state $\mathbf{x}_0$ is *stable* if for any small δ_1, there exists a δ_2 such that if $|\mathbf{x}(0) - \mathbf{x}_0| < \delta_2$ then the solution $\mathbf{x}(t)$ exists for all $t > 0$ and $|\mathbf{x}(t) - \mathbf{x}_0| < \delta_1$ for all $t > 0$. A stable steady state $\mathbf{x}_0$ is *asymptotically stable* if any solution $\mathbf{x}(t)$ with $\mathbf{x}(0)$ near $\mathbf{x}_0$ converges to $\mathbf{x}_0$ as $t \to \infty$. The steady state $\mathbf{x}_0$ is *unstable* if it is not stable.

Example 1. The equation

$$\frac{\mathrm{d}x}{\mathrm{d}t} = f(x) = x - x^3$$

has the steady states $x_0 = 0, \pm 1$. Note that $\mathrm{d}x/\mathrm{d}t > 0$ if $x < -1$ and $\mathrm{d}x/\mathrm{d}t < 0$ if $-1 < x < 0$. Hence $x(t) \to -1$ if $t \to \infty$ provided $x(0)$ is near -1, so that $x_0 = -1$ is asymptotically stable. Similar reasoning leads to the conclusion that $x_0 = +1$ is asymptotically stable and $x_0 = 0$ is unstable.

In the general case of eqn 3.14, if $\mathbf{x} = \mathbf{x}_0$ is a steady state, then one can write

$$\mathbf{f}(\mathbf{x}) = \mathbf{M}\boldsymbol{\xi} + o(|\boldsymbol{\xi}|), \tag{3.18}$$

where $\boldsymbol{\xi} = (\mathbf{x} - \mathbf{x}_0), o(|\boldsymbol{\xi}|) \to 0$ if $|\boldsymbol{\xi}| \to 0$, and $\mathbf{M}$ is the Jacobian matrix with elements

$$m_{ij} = \frac{\partial f_i}{\partial x_j}(\mathbf{x}_0). \tag{3.19}$$

The vector $\boldsymbol{\xi}$ denotes the deviation from the steady state $\mathbf{x}_0$. Consider the case $\mathbf{f}(\mathbf{x}) = \mathbf{M}\boldsymbol{\xi}$, then

$$\frac{\mathrm{d}\boldsymbol{\xi}}{\mathrm{d}t} = \mathbf{M}\boldsymbol{\xi}, \tag{3.20}$$

and consider a particular solution of the form $\boldsymbol{\xi}(t) = \boldsymbol{\xi}_0 e^{\lambda t}$; substituting this solution to eqn 3.20 leads to

$$\mathbf{M}\boldsymbol{\xi}_0 = \lambda \boldsymbol{\xi}_0. \tag{3.21}$$

This equation is called an eigenvalue equation, and $\boldsymbol{\xi}_0$ and λ are called *eigenvector* and *eigenvalue*, respectively. The eigenvalues of $\mathbf{M}$ are the roots $\lambda_1, \ldots, \lambda_n$ of the following nth-degree polynomial equation

$$\det(\lambda \mathbf{I} - \mathbf{M}) = \lambda^n + \alpha_1 \lambda^n + \cdots + \alpha_{n-1}\lambda + \alpha_n = 0. \tag{3.22}$$

If all the eigenvalues of $\mathbf{M}$ are different, then the general solution of eqn 3.20 is

$$\boldsymbol{\xi}(t) = \sum_{j=1}^{n} c_j \boldsymbol{\xi}_{0,j} e^{\lambda_j t}, \tag{3.23}$$

where the c_js are constants, and $\boldsymbol{\xi}_{0,j}$ is the eigenvector corresponding to the eigenvalue λ_j. It follows that if all the eigenvalues have negative real parts then $\boldsymbol{\xi}(t) \to \mathbf{0}$ (equivalently, $\mathbf{x}(t) \to \mathbf{x}_0$) as $t \to \infty$, so that $\mathbf{x}_0$ is asymptotically stable. If some of the eigenvalues coincide, say $\lambda_1 = \lambda_2 = \cdots = \lambda_k$, then one needs to replace c_j by $c_j t^{j-1} (j = 1, \ldots, k)$ to obtain the general solution. If at least one of the eigenvalues has a positive real part, then $\mathbf{x}_0$ is unstable; indeed, one can find an initial condition $\mathbf{x}(0)$ arbitrarily close to $\mathbf{x}_0$ such that $|\mathbf{x}(t)| \to \infty$ as $t \to \infty$. The next theorem deals with the general case where $\mathbf{f}(\mathbf{x})$ is given by eqn 3.18.

Theorem 3.3 *Let $\boldsymbol{\xi}$ be the deviation from the steady state $\mathbf{x}_0$ of the system of eqn 3.14 so that, by eqn 3.14, $\frac{\mathrm{d}\boldsymbol{\xi}}{\mathrm{d}t} = \mathbf{M}\boldsymbol{\xi} + o(|\boldsymbol{\xi}|)$ with $o(|\boldsymbol{\xi}|) \to 0$ if $|\boldsymbol{\xi}| \to 0$. If all the eigenvalues of $\mathbf{M}$ have negative real parts then $\mathbf{x}_0$ is asymptotically stable, and $|\boldsymbol{\xi}(t)| \leq (constant)e^{-\mu t}$ for some $\mu > 0$.*

3.3 Phase portraits on the plane

The set of all trajectories in phase space paints the *phase portrait* of the dynamical system. This portrait gives a global picture of the behavior of trajectories from all possible initial conditions or points in phase space. Consider the two-dimensional case of eqn 3.14 written explicitly as follows:

$$\begin{aligned}
\frac{\mathrm{d}x_1}{\mathrm{d}t} &= f_1(x_1, x_2) \\
\frac{\mathrm{d}x_2}{\mathrm{d}t} &= f_2(x_1, x_2).
\end{aligned} \tag{3.24}$$

Consider the case where $\mathbf{x}_0 = (x_1, x_2) = (0, 0)$ is a steady state, and, analogous to eqn 3.20, let the system linearized about $\mathbf{x}_0$ be the following:

$$\begin{aligned}
\frac{\mathrm{d}x_1}{\mathrm{d}t} &= m_{11}x_1 + m_{12}x_2 \\
\frac{\mathrm{d}x_2}{\mathrm{d}t} &= m_{21}x_1 + m_{22}x_2.
\end{aligned} \tag{3.25}$$

Figure 3.3 describes the behavior of trajectories near $\mathbf{x}_0$ when the eigenvalues λ_1, λ_2 are of the form:

(a) $\lambda_1 > 0, \lambda_2 > 0$;   (b) $\lambda_1 < 0, \lambda_2 < 0$;   (c) $\lambda_1 > 0, \lambda_2 < 0$;
(d) $\lambda_{1,2} = \alpha \pm i\beta, \alpha > 0$;   (e) $\lambda_{1,2} = \alpha \pm i\beta, \alpha < 0$;   and   (f) $\lambda_{1,2} = i\beta$.

To study the behavior of trajectories of eqn 3.24, it is useful to draw the *nullclines* of f_1 and f_2. The x_1-nullcline and the x_2-nullcline are defined by $f_1(x_1, x_2) = 0$ and $f_2(x_1, x_2) = 0$, respectively. A steady state of the system is a point where the two nullclines intersect. Figure 3.4 describes a situation where the two nullclines intersect at two points A and B.

As shown in Fig. 3.4, the phase plane is divided into five regions. In region I, $f_1 > 0$ and $f_2 < 0$ so that the vector field points toward the southeast. In region II, $f_1 < 0$ and $f_2 < 0$ so that the vector field points southwest, and so on. From the directions of

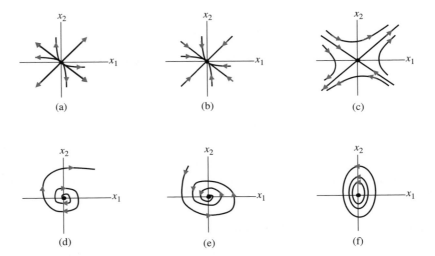

(a)   (b)   (c)

(d)   (e)   (f)

Fig. 3.3 Possible phase portraits on a plane according to the eigenvalues. Note that in case (f) the local trajectories are all periodic orbits.

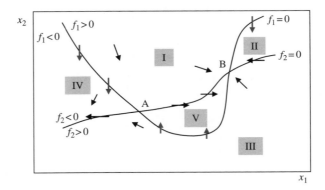

Fig. 3.4 Schematic illustration of nullclines (curves where $dx_1/dt = f_1 = 0$ and $dx_2/dt = f_2 = 0$) and directions of the vector field (*arrows*).

the vector field, it is seen that steady state B is stable and A is unstable. The phase diagram also shows that even if the initial condition $\mathbf{x}(0)$ is not near B, the solution $\mathbf{x}(t)$ may, nevertheless, converge to B as $t \to \infty$. For example, if $\mathbf{x}(0)$ belongs to region II or V then $\mathbf{x}(t) \to$ B as $t \to \infty$.

In Fig. 3.5 steady states A and C are stable, and steady state B is unstable. A system that has two stable steady states is said to be *bistable*. The system in Example 1 (Section 3.2.3) is bistable.

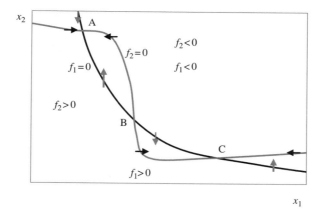

Fig. 3.5 Nullclines with three intersections corresponding to two stable steady states (A and C) and an unstable steady state (B). Directions of vector fields are shown by *arrows*.

3.4 Bifurcations

Experimentalists usually study their system of interest under controlled laboratory conditions. For example, the kinetics of biochemical reactions are often investigated under fixed temperature and pressure. In mathematics, these fixed conditions are referred to as the *parameters* of the system. Consider a system of ODEs that depends on a parameter p:

$$\frac{d\mathbf{x}}{dt} = \mathbf{f}(\mathbf{x}, p).$$ (3.26)

Bifurcation theory is concerned with the question of how solutions depend on the parameter p. For example, suppose that the steady state of eqn 3.26 depends on p, and that it is stable for $p < p_c$ but loses stability at p_c. A bifurcation at $p = p_c$ is said to have occurred, and p_c is referred to as a *bifurcation point*. Bifurcation points correspond to parameter values where a qualitative change occurs in the phase portrait of the system. Bifurcation is an important idea that will be useful in analyzing changes in the dynamics of the cellular processes considered in this book.

There are four major different types of bifurcations; the first three already occur in one-dimensional systems, while the fourth needs at least two dimensions. The first three types of bifurcations and their representative differential equations are as follows:

$$\frac{dx}{dt} = p + x^2 \quad \text{(saddle-point bifurcation)},$$ (3.27)

$$\frac{dx}{dt} = px - x^2 \quad \text{(transcritical bifurcation)},$$ (3.28)

$$\frac{dx}{dt} = px - x^3 \quad \text{(pitchfork bifurcation)}.$$ (3.29)

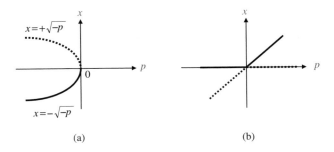

(a)   (b)

Fig. 3.6 (a) Saddle-point bifurcation diagram; (b) Transcritical bifurcation diagram.

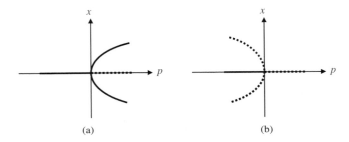

(a)   (b)

Fig. 3.7 Pitchfork bifurcations. *Solid curves* represent stable steady states, while *dotted curves* are unstable steady states.

In the case of eqn 3.27, there are two steady states for $p < 0$, namely, $x_{\pm}^s = \pm\sqrt{-p}$; x_{-}^s is stable and x_{+}^s is unstable. The solid curve in Fig. 3.6(a) describes the stable branch of steady states, and the dotted curve describes the curve of unstable steady states. At $p = 0$ the stable and unstable branches of steady states coalesce; this bifurcation at $p = 0$ is called a *saddle-point bifurcation*. Figure 3.6 gives examples of *bifurcation diagrams*. The parameter p is referred to as the *bifurcation parameter*.

In the case of eqn 3.28, $x = 0$ is a steady state for all p; it is stable if $p < 0$ and unstable if $p > 0$. Another branch of steady states is given by $x^s = p$; a steady state in this branch is stable for $p > 0$ and unstable for $p < 0$. This *transcritical bifurcation diagram* is shown in Fig. 3.6(b); the diagram is characterized by an exchange of stability of the branches of steady states at the bifurcation point (in this case, at $p = 0$).

In the case of eqn 3.29, for $p < 0$ the only steady state is $x^s = 0$; but for $p > 0$, there are, in addition to $x^s = 0$, two more steady states, namely, $x_{\pm}^s = \pm\sqrt{p}$. The steady state $x^s = 0$ is stable if $p < 0$ and unstable if $p > 0$. The steady states $x_{\pm}^s = \pm\sqrt{p}$ (for $p > 0$) are both stable. The diagram for this *pitchfork bifurcation* is shown in Fig. 3.7(a). A similar pitchfork bifurcation for the equation

$$\frac{dx}{dt} = px + x^3 \tag{3.30}$$

is shown in Fig. 3.7(b); in this case the two branches $x_{\pm}^{s} = \pm\sqrt{-p}$ are unstable and $x^{s} = 0$ is stable if $p < 0$ and unstable if $p > 0$. The bifurcation in Fig. 3.7(a) is said to be *supercritical* since the bifurcating branches appear for values of p larger than the bifurcation point $p_{c} = 0$. The bifurcation in Fig. 3.7(b) is called *subcritical* since the two bifurcating branches occur for p smaller than the bifurcation point $p_{c} = 0$.

3.5   Bistability and hysteresis

The pitchfork bifurcation diagram in Fig. 3.7(a), for $p > 0$, is an example of a *bistable* system characterized by having two stable steady states that coexist for a fixed value of p. As can be seen in Fig. 3.8, this pitchfork diagram is a 'slice' of the surface of steady states derived from the following equation:

$$\frac{\mathrm{d}x}{\mathrm{d}t} = x^{3} + \rho_{1}x + \rho_{0} = 0. \qquad (3.31)$$

The case of eqn 3.29 corresponds to $\rho_{0} = 0$ in eqn 3.31 – that is, the pitchfork bifurcation diagram is generated by the intersection of the surface defined by eqn 3.31 with the x-ρ_{1} plane.

A bifurcation diagram *distinguishes* one of the parameters of the system as *the* bifurcation parameter. Choosing ρ_{1} as the bifurcation parameter, one can then think of a bifurcation diagram as a 'slice' through the surface shown in Fig. 3.8 at a fixed value of the other parameter ρ_{0}. With this view, it is evident that the pitchfork diagram is

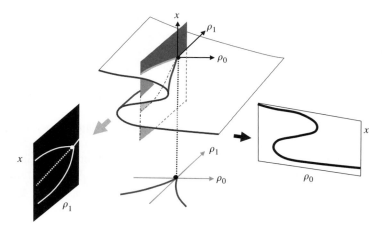

Fig. 3.8 The so-called cusp catastrophe manifold and its interactions with various planes to generate different bifurcation diagrams, such as the pitchfork diagram on the x-ρ_{1} plane (the *grey plane* shown on *lower left side*) and the Z-shaped diagram (*bold black curve* parallel to the x-ρ_{0} plane, shown on the *rightmost diagram*). The projection of the fold points of the manifold onto the ρ_{1}-ρ_{0} plane is the curve with a cusp at the origin.

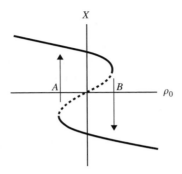

Fig. 3.9 Steady-state bifurcation diagram showing bistability in the range $A < \rho_0 < B$ (for $\rho_1 < 0$, see eqn (3.31). The *arrows* indicate components of the hysteresis loop. *Dotted curve* (middle branch) corresponds to unstable steady states.

unstable in the sense that small perturbations of the parameter ρ_0 break the pitchfork diagram into two disconnected curves.

The Z-shaped bifurcation diagram in Fig. 3.8 is redrawn in Fig. 3.9. The system is bistable in the parameter range $A < \rho_0 < B$, where A and B correspond to the knees of the curve. Using the method discussed in Section 3.2 one can show that the bottom and top branches of the curve are stable steady states, while the middle branch (dotted curve) are unstable steady states.

The experimental investigation of a bistable system requires varying a chosen bifurcation parameter (usually taken to be a condition that is amenable to experimental control – e.g. temperature, pressure, constant enzyme concentration). In the case shown in Fig. 3.9, if the parameter ρ_0 is increased (slowly enough for the system to settle to steady states, and therefore the experiment traces the curve of stable steady states) starting from the left, the value B will be reached where a discontinuous drop in the steady state is observed. If the experiment is repeated, but this time starting from a large value of ρ_0 and slowly decreasing it, one will notice that the system will pass through B (where the first jump was observed) but another value, this time A, is reached where the steady state jumps discontinuously. The loop formed by these two discontinuous transitions (including the segments of the curves of stable steady states) is referred to as a *hysteresis* loop. Thus, for values of ρ_0 within the bistable range $A < \rho_0 < B$, which of the two stable steady states of the system is reached depends on the initial condition.

3.6   Hopf bifurcation

Many biological processes are oscillatory in nature – the beating of the heart, spiking of neurons in the brain, circadian rhythms arising from cycles of day and night, the cell-division cycle, and many others (the book of Goldbeter (1996) provides many excellent examples). The existence and characteristics of these oscillations depend on the parameters of the system, and it is important to know when or how these oscillations arise. Hopf bifurcation refers to the bifurcation of periodic solutions as a parameter is varied.

Consider the system

$$\frac{dx}{dt} = f(x, y, p), \quad \frac{dy}{dt} = g(x, y, p). \tag{3.32}$$

Assume that for all p in some interval there exists a steady state $(x^s(p), y^s(p))$, and that the two eigenvalues of the Jacobian matrix (evaluated at the steady state) are complex numbers $\lambda_1(p) = \alpha(p) + i\beta(p)$ and $\lambda_2(p) = \alpha(p) - i\beta(p)$. Assume also that for some parameter p_0 within the interval the following are true:

$$\alpha(p_0) = 0, \quad \beta(p_0) \neq 0 \quad \text{and} \quad \frac{d\alpha}{dp}(p_0) \neq 0.$$

Then one of the following three cases must occur:

1. there is an interval $p_0 < p < c_1$ such that for any p in this interval there exists a unique periodic orbit containing $(x^s(p_0), y^s(p_0))$ in its interior and having a diameter proportional to $|p - p_0|^{1/2}$;
2. there is an interval $c_2 < p < p_0$ such that for any p in this interval there exists a unique periodic orbit as in case (1);
3. for $p = p_0$ there exist infinitely many orbits surrounding $(x^s(p_0), y^s(p_0))$ with diameters decreasing to zero (cf. Fig. 3.3(f)).

Case (1) is called a *supercritical Hopf bifurcation* and case (2) is called a *subcritical Hopf bifurcation*; these cases are generic. Case (3) is rather infrequent. Figure 3.10 illustrates a supercritical Hopf bifurcation.

The stability of the periodic solutions for cases (1) and (2) above is of interest. A periodic solution $\tilde{x}(t)$ is *stable* if any trajectory with initial condition $x(0)$ near $\tilde{x}(0)$ exists for all $t > 0$ and it converges to the periodic solution as $t \to \infty$. If this property is true only when $x(0)$ is outside (inside) the periodic solution then $\tilde{x}(t)$ is *stable from the outside (inside)*. Consider the case where the Jacobian matrix at the bifurcation point, p_0, is

$$\mathbf{M} = \begin{bmatrix} 0 & 1 \\ -b & 0 \end{bmatrix},$$

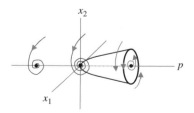

Fig. 3.10 Supercritical Hopf bifurcation at $p = 0$.

with eigenvalues $\pm bi$. Let W be defined by the expression below involving second-order and third-order derivatives (evaluated at $p = p_0$) of the functions f and g of the system of eqn 3.32,

$$W = b(f_{xxx} + f_{xyy} + g_{xxy} + g_{yyy})[f_{xy}(f_{xx} + f_{yy}) + g_{xy}(g_{xx} + g_{yy}) + f_{xx}g_{xx} - f_{yy}g_{yy}],$$
(3.33)

where $f_{xxx} = \frac{\partial^3 f}{\partial x^3}, g_{xxy} = \frac{\partial^3 f}{\partial x^2 \partial y}$, etc. The following give the criteria for stability of the periodic orbits (Guckenheimer and Holmes, 1983):

1. If $W < 0$ then the Hopf bifurcation is supercritical and the periodic solutions are stable.
2. If $W > 0$ then the Hopf bifurcation is subcritical and the periodic orbits are unstable.

■ 3.7 Singular perturbations

Biological processes often involve several time scales; an example would be the different time scales of gene expression and protein–protein interactions. Consider the following system with two time scales:

$$\frac{dx}{dt} = f(x, y),$$
(3.34)

$$\frac{dy}{dt} = \varepsilon g(x, y),$$
(3.35)

where ε is a small positive parameter; the variable y changes at a smaller scale than the variable x. Assume that the x-nullcline is cubic and the y-nullcline is a monotone function, and that these nullclines intersect at one point Q, as shown in Fig. 3.11. Then, there exists a periodic orbit of equations 3.34 and 3.35 which lies in the vicinity of the curve $ABCDA$. In order to explain why this is so, observe that if $\varepsilon = 0$ then $dy/dt = 0$ so that $y(t) = $ constant. Taking $\varepsilon = 0$ as an approximation is justified in case $f(x, y)$ stays away from zero in eqn 3.34 for then dy/dt is relatively negligible. Thus, a trajectory $(x(t), y(t))$ starting at (x_0, y_0) where $f(x_0, y_0) \neq 0$ will approximately satisfy $y(t) = y_0$ as long as $f(x(t), y_0)$ remains not equal to zero. In particular, a trajectory starting near and below B will travel horizontally, with speed $dx/dt = f > 0$, until it reaches the x-nullcline near C. Similarly, a trajectory starting near and above D will travel horizontally, with $dx/dt = f < 0$, until it reaches the x-nullcline near A.

To see how a trajectory evolves once it is located on the x-nullcline, one introduces the *slow time scale* τ defined as $\tau = \varepsilon t$. Then, eqns 3.34 and 3.35 become

$$\varepsilon \frac{dx}{d\tau} = f(x, y), \quad \frac{dy}{d\tau} = g(x, y).$$
(3.36)

Since $g < 0$ above the y-nullcline, g is negative along the arc AB and thus it pulls y further downward toward B as well as at B. Since B is the point at which the x-nullcline turns upward, the trajectory then departs from the x-nullcline, and then,

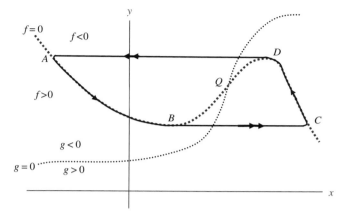

Fig. 3.11 Singular perturbation analysis of an oscillatory system. The curves corresponding to $f = 0$ and $g = 0$ are the nullclines of the system in eqns 3.34 and 3.35. See text for discussion.

as explained above, it proceeds along a curve close to the horizontal line BC. In a similar way, using the slow time scale, one can see that the trajectory will evolve approximately along the x-nullcline from C to D (since $dy/d\tau = g > 0$ on CD) and then back to A along $y = $ constant. The motion along the x-nullcline takes place in the slow time scale (slow motion), whereas the horizontal motion takes place on the original (fast) time scale as indicated by the double arrows in Fig. 3.11.

The above considerations can be made more precise. One can prove that for any small $\varepsilon > 0$ there exists a periodic orbit near the curve $ABCDA$. The above example is generic; it can be extended to the case of more than two ODEs showing, for example, the existence of periodic oscillations in networks.

3.8 Partial differential equations (PDEs)

3.8.1 Reaction-diffusion equations

Consider a particle that moves along a straight line, making steps of size Δx at discrete time intervals Δt. Denote by $p(x, t)$ the probability that the particle is at location x in time t. If the probability of stepping to the right is the same as the probability of stepping to the left, then

$$p(x, t + \Delta t) = \frac{1}{2}p(x + \Delta x, t) + \frac{1}{2}p(x - \Delta x, t). \tag{3.37}$$

If it is assumed further that Δx and Δt are small, then

$$p(x \pm \Delta x, t) = p(x, t) \pm p_x(x, t)\Delta x + \frac{1}{2}p_{xx}(x, t)(\Delta x)^2 + O(|\Delta x|^3),$$

$$p(x, t + \Delta t) = p(x, t) \pm p_t(x, t)\Delta t + O((\Delta t)^2),$$

where $O(|\Delta x|^3) \leq \text{const} \cdot |\Delta x|^3$, and $O((\Delta t)^2) \leq \text{const} \cdot (\Delta t)^2$. Substituting the above equations into eqn 3.37 and then letting $\Delta t \to 0$ and $(\Delta x)^2/(2\Delta t) \to \alpha$ (with $0 < \alpha < \infty$), one obtains

$$p_t = \alpha p_{xx}. \tag{3.38}$$

for $-\infty < x < \infty$ and $t > 0$.

Consider a chemical species with concentration $c(x,t)$. Interpreting the particle concentration at (x,t) as equivalent to the probability of finding the particle at (x,t), one has

$$c_t = \alpha c_{xx}. \tag{3.39}$$

Equation 3.39 is called the *diffusion equation,* and α is called the diffusion coefficient.

If the chemical species is involved in a set of chemical reactions whose net rate of producing this chemical species is $R(c)$, then one has the following *reaction-diffusion equation*

$$c_t = \alpha c_{xx} + R(c). \tag{3.40}$$

For example, if S and C are two chemical species with concentrations s and c, respectively, and if the set of reactions is $\{S \xrightarrow{k_1} C, C \xrightarrow{k_2} P\}$, then

$$c_t = \alpha c_{xx} + k_1 s - k_2 c, \tag{3.41}$$

where mass-action kinetics is assumed for both reactions. More generally, reaction-diffusion equations have the form

$$\frac{\partial c_i}{\partial t} = \nabla(D_i \nabla c_i) + R_i(c_1, \ldots, c_k; x) \quad \text{for } 1 \leq i \leq k, \tag{3.42}$$

where D_i is the diffusion coefficient of species i that may depend on spatial position and the concentrations of some species. The R_is are the rates of formation of c_i due to reactions at the point x.

■ 3.8.2 Cauchy problem

Consider the initial-value problem called the *Cauchy problem:*

$$\begin{aligned} c_t &= \alpha c_{xx} + f(x,t) \quad \text{for } -\infty < x < \infty, t > 0 \\ c(x,0) &= c_0(x) \quad \text{for } -\infty < x < \infty, \end{aligned} \tag{3.43}$$

where $f(x,t)$ and $c_0(x)$ are bounded continuous functions. Then, there is a unique bounded solution of eqn 3.43, and it is given by the following formula

$$c(x,t) = \int_{-\infty}^{\infty} K(x-y,t)c_0(y)dy - \int_0^t \int_{-\infty}^{\infty} K(x-y,t-\tau)f(y,\tau)dyd\tau, \tag{3.44}$$

where

$$K(x,t) = \frac{1}{\sqrt{4\pi\alpha t}}e^{-\frac{x^2}{4\alpha t}}. \tag{3.45}$$

$K(x,t)$ is called the *fundamental solution* of the diffusion equation.

The above consideration extends to n-dimensional space consisting of all points $\mathbf{x} = (x_1,\ldots,x_n)$, $-\infty < x_i < +\infty$ for $i = 1,\ldots,n$. Consider

$$\begin{aligned} c_t &= \alpha\nabla^2 c + f(\mathbf{x},t) && \text{for } \mathbf{x} \in R^n,\, t > 0 \\ c(\mathbf{x},0) &= c_0(\mathbf{x}) && \text{for } \mathbf{x} \in R^n, \end{aligned} \tag{3.46}$$

where $\nabla^2 \equiv \Delta \equiv \sum_{i=1}^{n} \frac{\partial^2}{\partial x_i^2}$ (called the *Laplacian*), and f and c_0 are bounded continuous functions. Then there exists a unique bounded solution of eqn 3.46, and it is given by

$$c(\mathbf{x},t) = \int_{R^n} K(\mathbf{x}-\mathbf{y},t)c_0(\mathbf{y})\mathrm{d}\mathbf{y} - \int_0^t \int_{R^n} K(\mathbf{x}-\mathbf{y},t-\tau)f(\mathbf{y},\tau)\mathrm{d}\mathbf{y}\mathrm{d}\tau, \tag{3.47}$$

where

$$K(\mathbf{x},t) = \frac{1}{(4\pi\alpha t)^{n/2}}e^{-\frac{|\mathbf{x}|^2}{4\alpha t}}. \tag{3.48}$$

If f and c in eqn 3.46 are independent of t, then eqn 3.46 becomes the *Poisson equation* $\alpha\nabla^2 c = f$; and if $f = 0$ then c satisfies the *Laplace equation* $\nabla^2 c = 0$.

■ 3.8.3 Dirichlet, Neumann and third-boundary-value problems

Consider the boundary-value problem

$$\begin{aligned} \nabla^2 u(x) &= f(x) && \text{for } x \in \Omega \\ \beta\frac{\partial u}{\partial n} + \gamma n &= g(x) && \text{for } x \in \partial\Omega, \end{aligned} \tag{3.49}$$

where Ω is a bounded domain (bounded, open and connected set) in R^n with boundary $\partial\Omega$, β and γ are given non-negative functions with $\beta^2 + g\gamma^2 > 0$, and f, g are given functions. The normal derivative $\partial/\partial n$ points in the direction outside of the domain Ω. The boundary-value problem is called the *Dirichlet problem* if $\beta \equiv 0$, $\gamma > 0$; it is called the *Neumann problem* if $\beta > 0$, $\gamma \equiv 0$, and the *third-boundary-value problem* if $\beta > 0, \gamma > 0$.

If f and g are continuous functions then the Dirichlet problem and the third-boundary-value problem have unique solutions. The uniqueness is a consequence of the *maximum principle* explained as follows. If

$$-\nabla^2 u(x) + b(x)\cdot\nabla u(x) + c(x)u(x) \geq 0 \quad \text{in } \Omega, \tag{3.50}$$

where $b(x)$, $c(x)$ are continuous functions and $c(x) \geq 0$, and if the maximum M of $u(x)$ in $\overline{\Omega} = \Omega \cup \partial\Omega$ is ≥ 0 and $u(x) \not\equiv M$ for some x in Ω, then the maximum cannot be taken at any point x_o in Ω; furthermore, if x_o is a point in $\partial\Omega$ where the maximum is obtained, then $\frac{\partial u}{\partial n}(x_o) > 0$, where n is the normal to $\partial\Omega$ pointing outward of Ω.

If the Neumann problem has a solution, then necessarily

$$\int_{\partial\Omega} \left(\frac{g}{\beta}\right) dS = \int_\Omega f d\Omega, \tag{3.51}$$

where dS and $d\Omega$ are surface and volume elements, respectively. When eqn 3.51 holds then the Neumann problem has a solution and, by the maximum principle, the solution is unique up to an additive constant.

Similarly to eqn 3.49, consider the initial-boundary-value problem

$$
\begin{aligned}
-u_t + \alpha\nabla^2 u &= f(x,t) && \text{for } x \in \Omega,\ t > 0 \\
\beta\frac{\partial u}{\partial n} + \gamma n &= g(x,t) && \text{for } x \in \partial\Omega,\ t > 0 \\
u(x,0) &= u_0(x) && \text{for } x \in \Omega,
\end{aligned}
\tag{3.52}
$$

where $\alpha > 0$. This problem has a unique solution provided $\beta > 0$, $\gamma \geq 0$, or $\beta \equiv 0$, $\gamma > 0$.

■ 3.9 Well posed and ill posed problems

In dealing with PDEs with initial and boundary conditions, and before proceeding with the calculation of the solution, one must determine whether the problem is *well posed*. The meaning of a well posed problem is explained below. Consider the example

$$
\begin{aligned}
\frac{\partial u}{\partial t} &= \Delta u && \text{for } x \in \Omega,\ t > 0 \\
u|_{\partial\Omega} &= 0 && \text{for } t > 0 \\
u(x,0) &= f(x) && \text{for } x \in \Omega,
\end{aligned}
\tag{3.53}
$$

where Ω is a bounded domain in R^n with boundary $\partial\Omega$. This problem has a unique solution for any continuous function $f(x)$. Furthermore, if

$$
\begin{aligned}
\frac{\partial v}{\partial t} &= \Delta v && \text{for } x \in \Omega,\ t > 0 \\
v|_{\partial\Omega} &= 0 && \text{for } t > 0 \\
v(x,0) &= g(x) && \text{for } x \in \Omega,
\end{aligned}
\tag{3.54}
$$

then the following is true: For any $T > 0$, and for arbitrarily small ε, $\max\limits_{\substack{x \in \bar{\Omega} \\ 0 \leq t \leq T}} |u(x,t) - v(x,t)| \leq \varepsilon$ provided the initial data satisfy $\max\limits_{x \in \bar{\Omega}} |f(x) - g(x)| \leq \delta$, where δ is sufficiently small. This property is referred to as the *stability* of the solution. When a problem has a unique solution (for any data in a 'large' class of functions) that also possesses the stability property, then the problem is *well posed*. A problem that is not well posed is said to be *ill posed*.

The following is an example of an ill posed problem:

$$
\begin{aligned}
\frac{\partial w}{\partial t} &= -\Delta w & \text{for } x \in \Omega, \quad t > 0 \\
w|_{\partial\Omega} &= 0 & \text{for } t > 0 \\
w(x,0) &= f(x) & \text{for } x \in \Omega.
\end{aligned}
\tag{3.55}
$$

Consider the simple case when $n = 1$ and $\Omega = \{x | \, 0 < x < 2\pi\}$. If

$$
f(x) = \sum_{m=1}^{M} a_m \sin mx,
\tag{3.56}
$$

then

$$
u(x,t) = \sum_{m=1}^{M} a_m e^{m^2 t} \sin mx,
\tag{3.57}
$$

which is a solution that defies the stability property if M is large enough. One can also show that the class of functions $f(x)$ for which there exists a solution to eqn 3.55 is very sparse in the class of continuous functions. (The functions $f(x)$ must be analytic, i.e. they can be expanded in a power series.)

It is important to note that biological problems may not always lead to well posed mathematical problems (for example, attempting to deduce from the present density of tumor cells their density at an earlier time).

3.10   Conservation laws

3.10.1   Conservation of mass equation

Consider a chemical species with density $\rho(x,t)$ at time t and position x, where $-\infty < x < \infty$, $t > 0$. Let the species move or flow in space with speed $v(x,t)$, that is,

$$
v(x,t) = \frac{dx}{dt}.
\tag{3.58}
$$

Consider a rectangle $x_1 < x < x_2$, $t_1 < t < t_2$. The change in mass of the chemical species in the interval $x_1 < x < x_2$ from times t_1 to t_2 is given by

$$
\int_{x_1}^{x_2} [\rho(x,t_2) - \rho(x,t_1)]dx.
$$

Equivalently, this mass change is due to the flows across $x = x_2$ and $x = x_1$, which gives the following net change in mass within the interval $x_1 < x < x_2$:

$$
\int_{t_1}^{t_2} [\rho v](x_1,t)dt - \int_{t_1}^{t_2} [\rho v](x_2,t)dt.
$$

If the chemical species is involved in processes that either create or consume it – with the net mass production rate $f(x,t)$ of the chemical species – then, by conservation of mass

$$\int_{x_1}^{x_2} [\rho(x,t_2) - \rho(x,t_1)]\mathrm{d}x = \int_{t_1}^{t_2} [\rho v](x_1,t)\mathrm{d}t - \int_{t_1}^{t_2} [\rho v](x_2,t)\mathrm{d}t + \int_{t_1}^{t_2}\int_{x_1}^{x_2} f(x,t)\mathrm{d}x\mathrm{d}t.$$

(3.59)

Using

$$\rho(x,t_2) - \rho(x,t_1) = (t_2 - t_1)\rho_t(x,t_1) + O(|t_2 - t_1|^2)$$
$$[\rho v](x_2,t) - [\rho v](x_1,t) = (x_2 - x_1)[\rho v]_x(x_1,t) + O(|x_2 - x_1|^2),$$

and dividing eqn 3.59 by $(t_2 - t_1)$ $(x_2 - x_1)$, and then letting $(t_2 - t_1) \to 0$ and $(x_2 - x_1) \to 0$, one obtains

$$\rho_t + (\rho v)_x = f,$$

(3.60)

which is the *conservation of mass equation*. This equation can be generalized to n-dimensions

$$\rho_t + \mathrm{div}(\rho \mathbf{v}) = f,$$

(3.61)

where $\mathbf{v}(\mathbf{x},t) = \mathrm{d}\mathbf{x}/\mathrm{d}t$, $\mathbf{x} \in R^n$.

3.10.2 Method of characteristics

Consider the initial-value problem of solving eqn 3.61 with the initial condition

$$\rho(\mathbf{x},0) = \rho_0(\mathbf{x}).$$

(3.62)

One method of solving this problem is the *method of characteristics*. Introduce the characteristic curves that end at $(\mathbf{x},\,t)$,

$$\frac{\mathrm{d}\xi_i}{\mathrm{d}\tau} = v_i(\xi,\tau), \quad 0 < \tau < t \quad (i = 1,\ldots,n),$$

(3.63)

$$\xi_i(t) = x_i \quad \text{(that is, } \boldsymbol{\xi}(t) = \mathbf{x}\text{)},$$

where $\mathbf{v}(\boldsymbol{\xi},\tau) = (v_1(\boldsymbol{\xi},\tau),\ldots,v_n(\boldsymbol{\xi},\tau))$, and denote the solution by $\boldsymbol{\xi}(\tau;\mathbf{x},t)$. Set $\mathbf{x}_0 = \boldsymbol{\xi}(0;\mathbf{x},t)$, $\bar{\rho}(\tau) = \rho(\boldsymbol{\xi}(\tau;\mathbf{x},t),\tau)$. Then

$$\frac{\mathrm{d}\bar{\rho}}{\mathrm{d}\tau} = \sum_{i=1}^n \frac{\partial\rho}{\partial\xi_i}\frac{\partial\xi_i}{\partial\tau} + \frac{\partial\rho}{\partial\tau} = \sum_{i=1}^n v_i\frac{\partial\rho}{\partial\xi_i} + \frac{\partial\rho}{\partial\tau} = \frac{\partial\rho}{\partial\tau} + \sum_{i=1}^n \frac{\partial}{\partial\xi_i}(\rho v_i) - \sum_{i=1}^n \bar{\rho}\frac{\partial v_i}{\partial\xi_i}, \quad (3.64)$$

or, by eqn 3.61,

$$\frac{d\bar{\rho}}{d\tau} = -\bar{\rho}\,\mathrm{div}(\mathbf{v}) + f \quad \text{for } 0 < \tau < t$$
$$\bar{\rho}(0) = \rho(\mathbf{x}_0). \tag{3.65}$$

Equation 3.65 is an ODE and its solution yields the solution of eqns 3.61 and 3.62:

$$\rho(\mathbf{x}, t) = \bar{\rho}(t) = \rho(\boldsymbol{\xi}(0; \mathbf{x}, t), 0). \tag{3.66}$$

In the case where the function $\rho(\mathbf{x}, t)$ is restricted to $\mathbf{x} \in \Omega$, where Ω is a bounded domain, the characteristic curves $\boldsymbol{\xi}(t)$ from points $(\mathbf{x}, t)$ may reach the boundary $\partial\Omega$ at some $\tau > 0$; in this situation, boundary values must be prescribed at these 'exit points' of $\partial\Omega$. For simplicity, consider first the case

$$\frac{\partial\rho}{\partial t} + \alpha\rho = 0 \quad \text{for } 0 < x < A, \tag{3.67}$$

where α is a positive constant. Then, the characteristic curves are defined by

$$\frac{d\xi}{d\tau} = \alpha \text{ or } \xi(\tau) = (x - \alpha t) + \alpha\tau \quad \text{(with } \xi(t) = x). \tag{3.68}$$

If $x > \alpha t$ then the characteristic curve does not exit the interval $(0, A)$ for all $0 < \tau < t$; but if $x < \alpha t$ then it exits this interval at $\xi = 0$ at time $\tau = t - x/\alpha$. Thus, initial and boundary values must be assigned:

$$\rho(x, 0) = \rho_0 \qquad \text{for } 0 < x < A, \tag{3.69}$$
$$\rho(0, t) = \rho_1(t) \quad \text{for } 0 < t < \infty. \tag{3.70}$$

Consider a more general case:

$$\frac{\partial\rho}{\partial t} + \mathrm{div}(\rho\mathbf{v}) = F(\rho, x) \quad \text{for } x \in \Omega, \ t > 0$$
$$\rho(x, 0) = \rho_0(x), \qquad x \in \Omega, \tag{3.71}$$

where $\Omega = \{(x_1, x_2) | 0 < x_1 < A, \ 0 < x_2 < A\}$. Furthermore, assume that

$$v_1(0, x_2, t) > 0, \ v_1(A, x_2, t) < 0$$
$$v_2(x_1, 0, t) > 0, \ v_2(x_1, A, t) < 0. \tag{3.72}$$

Then, the characteristic curves $\boldsymbol{\xi}(t)$ $(\tau < t)$ with $\boldsymbol{\xi}(t) = x \in \Omega$ can exit Ω at any point of the boundary $\partial\Omega$. Hence, one needs to prescribe

$$\rho(x, t) = \rho_1(x, t) \quad \text{for } x \in \partial\Omega, \ t > 0, \tag{3.73}$$

after which eqns 3.71–3.73 can then be solved using the method of characteristics.

3.11   Stochastic simulations

Chemical reactions occur by random or stochastic collisions between reactant molecules. It was demonstrated by Oppenheim *et al.* (1969) and proved rigorously by Kurtz (1972) that these stochastic processes are described exactly by the deterministic eqn 3.5 – for spatially homogeneous sytems – in the thermodynamic limit, that is, when the number of molecules of each species approaches infinity and the spatial domain increases to the entire space in such a way that molecular concentrations approach a finite limit. However, when the number of molecules becomes small, the deterministic equations are no longer appropriate.

Gillespie (1976) developed an algorithm for simulating stochastic chemical reactions. The starting point is a molecular mechanism composed of elementary steps (involving no intermediate steps) in the conversion of reactants to products. These elementary steps are usually unimolecular or bimolecular. The *propensity* of each elementary step is defined according to the probability of collision among the reactants. As examples, the propensity (symbolized by a_v) for the unimolecular reaction $\{R \xrightarrow{k_1} P\}$ is $a_1 = k_1 X_R$, while that of a bimolecular reaction $\{A + B \xrightarrow{k_2} C\}$ is equal to $a_2 = k_1 X_A X_B$, where the Xs refer to molecule numbers and the ks to some constant parameters. Note that for the reaction $2S \xrightarrow{k_3} P$, the propensity is not $k_3 X_S^2$ but, instead, $a_3 = k_3 X_S (X_S - 1)/2$ because one has to avoid overcounting pairwise collisions. The Gillespie algorithm is summarized in Fig. 3.12.

The algorithm has two stochastic steps employing random numbers r_1 and r_2 in the interval $[0, 1]$. First, the probability at time t that a reaction occurs in the time interval between $t + \tau$ and $t + \tau + d\tau$ is given by the following Poisson distribution

$$P(t, \tau) = a_0 \exp(-a_0 \tau), \quad \text{where } a_0(t) = \sum_{v=1}^{r} a_v(\mathbf{X}(t)). \tag{3.74}$$

As shown in Fig. 3.12, given the numbers of all molecules and the rate parameters k_v, the sum of all the reaction propensities, a_0, is calculated and the time τ for the next reaction is calculated using the random number r_1 from

$$\tau = (1/a_0) \ln(1/r_1). \tag{3.75}$$

Comparison with eqn 3.74 shows that the underlying hypothesis here is that $P(t, \tau)$ takes a random value r_1.

Second, which reaction μ among the r reactions is chosen to occur at time τ is decided by the relative values of the propensities of the reactions, as illustrated in Fig. 3.12. The larger a_μ/a_0 is, the larger is the probability for r_2 to fall in the interval $\sum_{v=1}^{\mu-1} (a_v/a_0) < r_2 \leq \sum_{v=1}^{\mu} (a_v/a_0)$ and thus for the reaction μ to occur. Finally, time is increased by τ and the number of molecules is updated according to the stoichiometry of reaction μ.

The Gillespie algorithm is essentially an exact procedure for numerically simulating the time evolution of a well-stirred chemically reacting system. Its drawback is that it

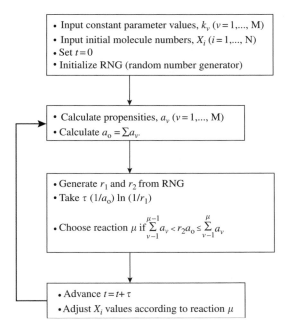

Fig. 3.12 The Gillespie stochastic algorithm (Gillespie, 1977). See text for explanation.

requires a large amount of computer time. Significant gains in simulation speed with some loss in accuracy are described by Gillespie (2001).

3.12 Computer software platforms for cell modelling

The systems biology community has been quite active in developing computer software platforms for simulating biochemical reaction networks in a cell. A listing of the more popular free software on the internet is given in Table 3.1.

CellDesigner, VCell, and E-Cell are 'whole-cell' simulation platforms that allow the user to consider subcellular localizations (cytoplasm, nucleus, mitochondria, etc.) of molecules and simulate the dynamics of biochemical reactions. CellDesigner is perhaps the most developed and most advanced in integrating standards of systems biology and bioinformatics. The program has a good facility for drawing gene-regulatory and biochemical networks, for generating the differential equations associated with a model network, and for performing dynamic simulations of the model. XPPAUT (or simply XPP) is a popular software for solving differential equations, difference equations, delay equations, functional equations, boundary-value problems, and stochastic equations. The 'AUT' in the name of the software refers to a program called AUTO that carries out bifurcation analysis of differential equations.

Table 3.1 Some popular software platforms for modelling cellular networks.

Name of software platform	Internet address
CellDesigner	http://celldesigner.org
VCell	http://www.nrcam.uchc.edu/
CellML	http://www.cellml.org/
E-Cell	http://www.e-cell.org/
JDesigner	http://sbw.kgi.edu/software/jdesigner.htm
Copasi	http://www.copasi.org/tiki-index.php
XPPAUT	http://www.math.pitt.edu/~bard/xpp/xpp.html

References

Clarke, B. L. (1980) 'Stability of Complex Reaction Networks', *Advances in Chemical Physics* **43**, 1–215.

Feinberg, M. (1980) 'Lectures on Chemical Reaction Networks.' *Technical Report of the Mathematics Research Center*, University of Wisconsin-Madison.

Gillespie, D. T. (1976) 'A general method for numerically simulating the stochastic time evolution of coupled chemical reactions,' *Journal of Computational Physics* **22**, 403–434.

Gillespie, D. T. (1977) 'Exact stochastic simulation of coupled chemical reactions', *Journal of Physics Chemistry* **81**, 2340–2361.

Gillespie, D. T. (2001) 'Approximate accelerated stochastic simulation of chemically reacting systems,' *Journal of Chemical Physics* **115**, 1716–1733.

Goldbeter, A. (1996) Biochemical oscillations and cellular rhythms. Cambridge University Press.

Guckenheimer, J. and Holmes, P. (1983). *Nonlinear oscillations, dynamical systems, and bifurcation of vector fields*, Springer Verlag, New York.

Kurtz, T. G. (1972)'The relationship between stochastic and deterministic models for chemical reactions,' *Chemical Physics* **57**, 2976–2978.

Oppenheim, I., Shuler, K. E., and Weiss, G. H. (1969) 'Stochastic and deterministic formulation of chemical rate equations,' *Journal of Chemical Physics* **50**, 460–466.

Exercises

1. Using the steady-state approximation (that is, $d[E]/dt = 0$ or $d[ES]/dt = 0$) and setting $v = d[P]/dt$, derive eqn 3.7, and show that $V_{max} = k_2 E_{tot}$ and $K_M = (k_{-1} + k_2)/k_1$.
2. Using the steady-state approximation for $[ES]$ and $[ES_2]$ for the mechanism in eqn 3.10, show that the steady-state rate of formation of the product is given by eqn 3.11.
3. Derive eqn 3.13 and give the expressions for V_{max} and K_M^n.

4. Prove Theorem 3.1. Write eqns 3.1 and 3.2 in the form

$$\mathbf{x}(t) = \mathbf{x}_0 + \int_0^t \mathbf{f}(\mathbf{x}(s))ds. \qquad (*)$$

Hint: Define

$$\mathbf{x}_{m+1}(t) = \mathbf{x}_0 + \int_0^t \mathbf{f}(\mathbf{x}_m(s))ds, \quad \mathbf{x}_0(t) \equiv \mathbf{x}_0.$$

Show by induction on m that

$$|\mathbf{x}_{m+1}(t) - \mathbf{x}_m(t)| \le \frac{ct^m}{m!}$$

and that $\lim_{m\to\infty} \mathbf{x}_m$ is a solution asserted in Theorem 3.1.
5. Prove that the solution of (*) in Exercise 4 is unique.
6. Prove Theorem 3.2.
7. Prove Theorem 3.3.
8. Determine whether the steady states of the following systems are stable or not:

(a) $\quad \dfrac{dx}{dt} = x + y - x^2, \quad \dfrac{dy}{dt} = x - y$

(b) $\quad \dfrac{dx}{dt} = y - \gamma x + x^2, \quad \dfrac{dy}{dt} = x - y - 1 \quad (\gamma > 0).$

9. Show that the cubic polynomial $x^3 + \beta_2 x^2 + +\beta_1 x + \beta_0$ can be reduced to the form of the right-hand side of eqn 3.31 by applying a translation of the variable x.
10. Consider the following systems

(a) $\quad \dfrac{dx}{dt} = y, \quad \dfrac{dy}{dt} = -y^n + py - x \quad (n = 1, 2, 3, \ldots),$

(b) $\quad \dfrac{dx}{dt} = -x + x^2, \quad \dfrac{dy}{dt} = x + y,$

(c) $\quad \dfrac{dx}{dt} = y, \quad \dfrac{dy}{dt} = py - x - x^2 - x^3.$

For each system above, at which value of p does bifurcation occur? What is the type of bifurcation?
11. Prove that (a) the function 3.48 satisfies the diffusion equation $K_t = \nabla^2 K$, (b) the function 3.47 with $f \equiv 0$ satisfies eqn 3.46 with $f \equiv 0$.

4

Gene-regulatory networks: from DNA to metabolites and back

DNA has at least three functions essential to life: first, its chemistry provides a mechanism for replicating genes (via the Watson–Crick base pairing); second, in its long strands are nucleotide sequences containing information for producing proteins (using the universal genetic code); and, third, DNA provides instructions on how a cell would respond or adapt to its environment for growth and survival. The third function will be illustrated in this chapter using specific examples of gene-regulatory networks (GRNs). A living cell possesses a high degree of autonomy – which means that as long as raw materials are available, it is able to synthesize most of what it needs. (For example, the human body can synthesize 12 of the 20 amino acids found in cellular proteins; the remaining 8, called *essential amino acids*, must be included in the diet.) GRNs that act as sensors and controllers of cellular responses to levels of metabolites in their environments have been elucidated. Two of these, in the bacterium *Escherichia coli*, are discussed in detail in this chapter. The first GRN involves the amino acid tryptophan, and the second is on the utilization of lactose. These models demonstrate the connectedness of GRNs – from DNA to RNA to proteins to metabolites and back. With regards to modelling, an important lesson illustrated in this chapter is the process of creating low-dimensional models from complex GRNs.

4.1 Genome structure of *Escherichia coli*

A large number of prokaryotic genes are organized in chromosomal blocks called *operons*. An operon is composed of contiguous genes that are transcribed as single mRNA. Transcription of an operon is controlled by a promoter-operator region on the DNA where transcription factors (usually proteins) bind to regulate the rate of transcription initiation. Figure 4.1(a) shows the relative positions of some operons in the circular chromosome of *E. coli*. Most operon genes code for proteins that collaborate in implementing a specific cellular function; for example, the Trp operon (Fig. 4.1(b)) in *E. coli* is composed of genes involved in the synthesis of the amino acid tryptophan. In contrast, in eukaryotes, the corresponding genes for Trp synthesis do not form an operon; for example, in the yeast *S. cerevisiae* the genes are located in four different linear chromosomes (see Fig. 4.1(c)).

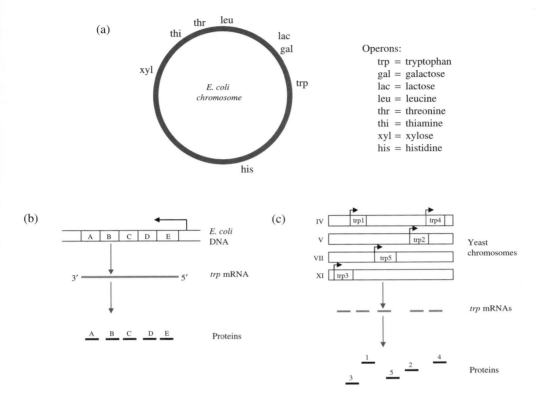

Fig. 4.1 (a) Relative positions (schematic) of a few of the operons in the circular chromosome of *E. coli* (there are over 400 operons known). (b) Trp operon in *E. coli* showing the component genes *trpA* to *trpE*. (c) The corresponding genes in yeast (*S. cerevisiae*) are not organized into an operon but are found in four different linear chromosomes (IV, V, VII, and XI). Figures (b) and (c) are adapted from Fig. 4–17 in the textbook of Lodish *et al.* (1999).

4.2 The Trp operon

The genes comprising the Trp operon, the network regulating their expression, and the metabolic pathway for Trp synthesis are shown in Fig. 4.2. The structural genes *trpA* to *trpE* form a single block on the DNA (shown as *trpABCDE* in the figure). There are three important negative-feedback mechanisms in the network: transcriptional repression, end-product enzyme inhibition, and transcriptional attenuation (labelled with the encircled numbers 1 to 3, respectively).

Transcriptional repression occurs at the operator region *trpO* that has some overlap with the promoter region *trpP*. A repressor protein, R, binds *trpO*, thereby preventing the RNA polymerase from binding to *trpP*. The repressor gene *trpR* is found near the promoter-operator region. The repressor protein exists as a dimer, with each monomer containing a binding site for Trp. Without bound Trp molecules, the repressor cannot

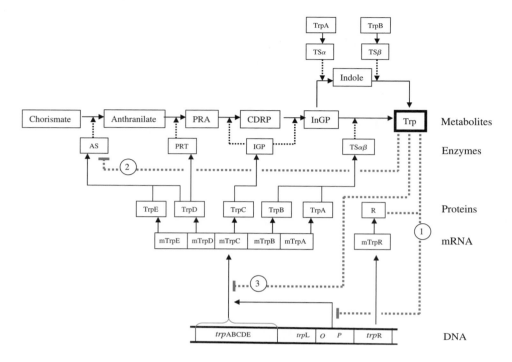

Fig. 4.2 The Trp operon regulatory network. Abbreviations: Trp = tryptophan, PRA = N-(5′-phosphoribosyl)-anthranilate, CDRP = 1-(o-carboxyphenylamino)-1′-deoxyribulose-5′-phosphate, InGP = indole-3-glycerol-phosphate, AS = anthranilate synthase, PRT = phosphoribosyl anthranilate transferase, IGP = indole-3-glycerol phosphate synthase, TSα = tryptophan synthase (α subunit), TSβ = tryptophan synthase (β subunit), TSαβ = tryptophan synthase (α and β subunits), R = repressor.

bind tightly to *trpO* and does not inhibit transcription; this inactive form of the repressor is called the *aporepressor*.

End-product enzyme inhibition in the network of Fig. 4.2 involves the product, Trp, inhibiting the first enzyme, AS (anthranilate synthase), of the metabolic pathway that leads to the synthesis of Trp itself. The functional enzyme AS is a heterotetramer consisting of two TrpE and two TrpD proteins. Binding of Trp molecules to the TrpE subunits inactivates AS.

Transcriptional attenuation is a subtle regulatory mechanism, and is the weakest among the three negative feedbacks. The transcript of the leader region *trpL* (see Fig. 4.2) has 4 segments (labelled 1, 2, 3, 4) that can form three stable hairpin structures 1:2, 2:3, and 3:4. Nucleotide sequence analysis shows that segment 1 includes two Trp codons in tandem. The 1:2 hairpin, closest to *trpO*, causes a pause in transcription that gives time for a ribosome to bind the leader transcript and start translation. The movement of the ribosome disrupts the 1:2 hairpin and resumes transcription.

High Trp levels lead to abundant charged tRNA[Trp] molecules causing ribosomes to initiate translation of the leader segments more rapidly. The probability of forming the 3:4 hairpin structure is thus increased compared to the 1:2 or 2:3 hairpins. Interestingly, the RNA polymerase recognizes the 3:4 structure as a signal for terminating transcription. On the other hand, when Trp levels are low, a ribosome could stall at either of the Trp codons because of small numbers of charged tRNA[Trp]. The 2:3 hairpin structure is thus formed instead of the 1:2 structure, and formation of the 3:4 structure that terminates transcription is prevented. The 1:2 structure is often referred to as the antiterminator because its formation allows transcription to the end of the operon.

4.3 A model of the Trp operon

Santillan and Zeron (2004) formulated a model of the Trp-operon dynamics incorporating the three negative-feedback mechanisms discussed in the previous section. The Santillan–Zeron (SZ) model has three variables M, E_{tot}, and T_{tot}, where $M = trpE$ mRNA concentration, $E_{tot} = $ total AS concentration, and $T_{tot} = $ total Trp concentration. The model equations – whose derivations will be discussed below – are the following:

$$\frac{dM}{dt} = k_m O_{tot} \frac{\frac{P}{K_P}}{1 + \frac{P}{K_P} + \frac{R_{tot}}{K_R}\left(\frac{T}{K_T + T}\right)^2} \times \frac{1 + \frac{2\alpha T}{K_G + T}}{(1 + \frac{\alpha T}{K_G + T})^2} - (\gamma_M + \mu)M, \qquad (4.1)$$

$$\frac{dE_{tot}}{dt} = \frac{1}{2} k_E M(t - \tau_E) - (\gamma_E + \mu)E_{tot}, \qquad (4.2)$$

$$\frac{dT_{tot}}{dt} = k_T E_{tot} \left(\frac{K_I}{K_I + T}\right)^2 - \rho \frac{T}{K_\rho + T} - \mu T_{tot}, \qquad (4.3)$$

where $M(t - \tau_E)$ in eqn 4.2 is the concentration M delayed by time τ_E, taken to be the time to translate the TrpE protein. The other symbols in the above equations are parameters of the model; their definitions and values are given in Table 4.1.

Equations 4.1–4.3 are derived using various simplifying assumptions. The active repressor (R_{2T}) competes against the RNA polymerase (P) for available promoter-operator regions (O):

$$R_{2T} + O \xrightarrow{K_R} O_R, \qquad (4.4)$$

$$P + O \xrightarrow{K_P} O_P, \qquad (4.5)$$

where O_R and O_P are the O–R_{2T} and O–P complexes, respectively. K_R and K_P are the respective equilibrium constants for the dissociation of these complexes (double-headed arrows indicate reversibility of the steps); each equilibrium constant is defined as the ratio of the rate constant of the dissociation of the complex over the rate constant of the association reaction. Assuming that reactions 4.4 and 4.5 rapidly attain

Table 4.1 Parameters of the Santillan–Zeron model of the Trp operon.

Parameter symbol	Description	Values used by Santillan and Zeron (2004)
O_{tot}	Total operator-promoter concentration	$4 \times 10^{-3}\,\mu\text{M}$
R_{tot}	Total repressor concentration	$0.8\ \mu\text{M}$
P	RNA polymerase concentration	$3.0\ \mu\text{M}$
K_T	equilibrium constant of dissociation of repressor–tryptophan complex	$40\ \mu\text{M}$
K_R	equilibrium constant of dissociation of operator–repressor complex	$2 \times 10^{-4}\,\mu\text{M}$
K_P	equilibrium constant of dissociation of operator–polymerase complex	$4.5 \times 10^{-2}\,\mu\text{M}$
K_G	equilibrium dissociation constant of Trp-charged tRNA	$5\ \mu\text{M}$
K_I	equilibrium constant of dissociation of tryptophan–AS enzyme complex	$4.1\ \mu\text{M}$
K_ρ	tryptophan consumption parameter	$10\ \mu\text{M}$
k_m	rate coefficient of transcription of operon	$5.1\ \text{min}^{-1}$
k_E	rate coefficient of synthesis of AS enzyme	$30\ \text{min}^{-1}$
k_T	rate coefficient for tryptophan synthesis	$7.3 \times 10^4\ \text{min}^{-1}$
μ	growth rate coefficient of bacterium	$10^{-2}\ \text{min}^{-1}$
α	transcription attenuation parameter	18.5
γ_M	degradation rate coefficient of *trpE* mRNA	$0.69\ \text{min}^{-1}$
γ_E	degradation rate coefficient of AS	$0\ \text{min}^{-1}$
ρ	rate coefficient for tryptophan consumption	$2.4 \times 10^2\,\mu\text{M min}^{-1}$
τ_E	time to translate a TrpE protein	$1\ \text{min}$

equilibrium, the following relationships are valid:

$$K_R O_R = O R_{2T} \quad \text{and} \quad K_P O_P = OP, \tag{4.6}$$

where the italicized symbols of the chemical species represent their respective concentrations. In the SZ model, the total concentration of the operator-promoter regions, O_{tot}, is assumed constant. Since $O_{\text{tot}} = O + O_R + O_P$, using eqn 4.6, one gets

$$O_P = O_{\text{tot}} \frac{\frac{P}{K_P}}{1 + \frac{P}{K_P} + \frac{R_{2T}}{K_R}}. \tag{4.7}$$

The concentration of the active repressor, R_{2T}, depends on the total concentration of the aporepressor (R_{tot}, assumed constant) and the Trp concentration (T), as can be expected from the sequential binding reactions of Trp (T) to the aporepressor (R):

$$\text{R} + \text{T} \overset{K_T/2}{\longleftrightarrow} \text{R}_\text{T} \quad \text{and} \quad \text{R}_\text{T} + \text{T} \overset{2K_T}{\longleftrightarrow} \text{R}_\text{2T}, \tag{4.8}$$

where R_T is the single-Trp-bound repressor, R_{2T} is the double-Trp-bound repressor, and K_T is the equilibrium constant for the dissociation of the repressor–Trp complex defined by

$$\frac{1}{2}K_T R_T = RT \quad \text{and} \quad 2K_T R_{2T} = R_T T. \tag{4.9}$$

In eqns 4.8 and 4.9, the factor $1/2$ occurs because the first T that binds R has two available binding sites; and the factor 2 accounts for the dissociation rate of R_{2T} being twice the dissociation rate of a single T from the complex. In the Santillan–Zeron (SZ) model, the total repressor concentration, R_{tot} (where $R_{\text{tot}} = R + R_T + R_{2T}$), is assumed constant. From eqn 4.9, one obtains the following expression:

$$R_{2T} = R_{\text{tot}} \left(\frac{T}{K_T + T} \right)^2. \tag{4.10}$$

The rate of transcription of the operon is proportional to the concentration of the O_P (promoter–RNA polymerase) complex; this rate has the following constant value derived from eqns 4.7 and 4.10:

$$k_m O_P = k_m O_{\text{tot}} \frac{\frac{P}{K_P}}{1 + \frac{P}{K_P} + \frac{R_{\text{tot}}}{K_R} \left(\frac{T}{K_T + T} \right)^2}. \tag{4.11}$$

Recall that the mechanism of transcription attenuation, depending on Trp concentration, can terminate transcription. Santillan and Zeron (2004) did not consider all the steps of the attenuation mechanism but derived the probability that transcription is not terminated – this is equal to the probability that a ribosome translating the leader mRNA stalls at either of the Trp codons in segment 1:

probability that transcription is not terminated by attenuation

$$= \frac{1 + \frac{2\alpha T}{K_G + T}}{(1 + \frac{\alpha T}{K_G + T})^2}, \tag{4.12}$$

where K_G is the equilibrium constant for the dissociation of charged tRNA^{Trp}, and $\alpha = G_{\text{tot}}/K_G$ (where $G_{\text{tot}} = $ constant total tRNA^{Trp} concentration).

Combining eqns 4.11 and 4.12 gives the first term on the right-hand side of eqn 4.1. The second term of this equation represents the degradation rate of *trpE* mRNA (with rate coefficient γ_M) and rate of dilution of *trpE* mRNA due to bacterial growth (with coefficient μ).

The step catalyzed by the enzyme AS is known to be the slowest and rate-determining step of the metabolic pathway for Trp synthesis from chorismate – this is the justification for focusing only on the dynamics of this enzyme. The rate of synthesis of AS is proportional to the concentration of *trpE* mRNA that, when the time τ_E to completely translate a TrpE protein is considered, is equal to $\frac{1}{2}k_E M(t - \tau_E)$, as shown in the first term of the right-hand side of eqn 4.2. The factor $1/2$ occurs because two TrpE proteins are required to assemble one AS molecule (see Exercise

2 for justification). The second term on the right-hand side of eqn 4.2 represents the degradation rate of AS (with rate coefficient γ_E) and the dilution rate due to bacterial growth (with rate coefficient μ).

The binding steps of tryptophan to AS are represented by the following reversible processes:

$$E + T \xleftrightarrow{K_I/2} E_T \quad \text{and} \quad E_T + T \xleftrightarrow{2K_I} E_{2T}. \tag{4.13}$$

where E is the free enzyme, E_T is single-Trp-bound enzyme, and E_{2T} is the double-Trp-bound enzyme. These steps are analogous to those of eqn 4.8. K_I is the equilibrium constant for the dissociation of the enzyme–tryptophan complex defined by

$$\frac{1}{2}K_I E_T = ET \quad \text{and} \quad 2K_I E_{2T} = E_T T. \tag{4.14}$$

The total enzyme concentration, E_{tot}, is given by

$$E_{tot} = E + E_T + E_{2T}. \tag{4.15}$$

Ignoring the contributions of the repressor–Trp complexes (since the total repressor concentration is negligible compared to those of Trp and AS), the total Trp concentration is given by

$$T_{tot} = T + E_T + 2E_{2T}. \tag{4.16}$$

From eqns 4.14–4.16, one obtains the following expression for the free enzyme and free Trp concentrations:

$$E = E_{tot}\left(\frac{K_I}{K_I + T}\right)^2, \tag{4.17}$$

$$T = \frac{1}{2}\sqrt{(K_I + 2E_{tot} - T_{tot})^2 + 4K_I T_{tot}} - \frac{1}{2}(K_I + 2E_{tot} - T_{tot}). \tag{4.18}$$

Equation 4.3 for the dynamics of total Trp concentration assumes that the rate of Trp synthesis is proportional to the concentration of the free enzyme (eqn 4.17). The second term on the right-hand side of eqn 4.3 stands for the Trp consumption rate, and the last term is the dilution rate of Trp due to bacterial growth; this completes the derivation of the set of the model in eqns 4.1–4.3.

4.4 Roles of the negative feedbacks in the Trp operon

In a Trp-rich culture medium, the *E. coli* bacterium does not need to synthesize Trp. The inhibition of AS by Trp is energy efficient because this enzyme is at the start of the Trp-synthesis pathway; inhibiting any other enzyme in the pathway would be inefficient. Furthermore, targeting an enzyme in the metabolic pathway is a quicker mechanism for sensing increases in Trp compared to targetting the transcription of

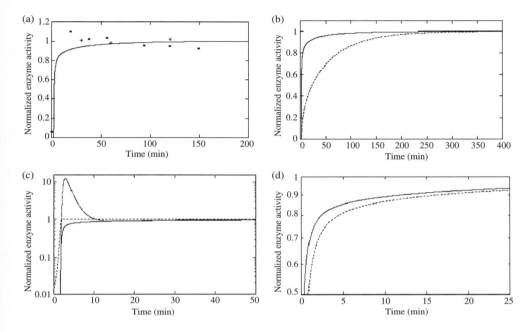

Fig. 4.3 Activity of the enzyme AS (anthranilate synthase) normalized over total enzyme. (a) Derepression experiment (see text for discussion); model simulation (*solid curve*) compared to experimental data (*crosses and squares*). (b) Derepression dynamics of free AS (*solid curve*) and total AS (*dashed curve*). (c) Derepression dynamics of free AS in a normal Trp operon (*solid line*), when AS inhibition is the only negative feedback (*dashed curve*), and when AS inhibition is absent (*dot-dashed line*). (d) Dynamics of free AS in a reactivation experiment (see text for discussion) with a normal Trp operon (*solid curve*) and with an attenuation-lacking operon (*dashed curve*). Figures are reproduced with permission from the paper of Santillan and Zeron (2004). Copyright 2004 Elsevier Ltd.

the Trp operon genes. But under conditions of sustained high Trp levels, repression of gene expression would make sense because it would be wasteful otherwise. However, shutting off operon expression completely could become problematic when Trp levels become so low that Trp-containing proteins can no longer be produced. So, in fact, even in Trp-rich media low levels of operon expression do occur, and it is this that prepares the cell for rapid synthesis of the amino acid when the need arises.

Computer simulations using the SZ model offer detailed insights into the dynamics of the Trp operon. Figure 4.3(a) demonstrates that the SZ model with the parameter values given in Table 4.1 is able to reproduce observations from derepression experiments. These experiments involve growing *E. coli* in Trp-rich medium for a long time to ensure that the Trp operon is fully repressed; derepression of the operon is then carried out by transferring the bacteria in a Trp-free medium, and enzyme activities are then measured at different times. In Fig. 4.3(a) experimental measurements of the activity of the enzyme AS are compared with the results of model simulations. The experiments show that steady-state AS activity is attained rapidly (in about 20 min); this quick response is reproduced by the SZ model. Figure 4.3(b) is more revealing; this figure shows that in a derepression experiment, the activity of AS – interpreted in the model to be the free AS with concentration E given by eqn 4.17 – approaches steady state more quickly than the total E_{tot} (dashed curve in Fig. 4.3(b)). Thus, Trp is acting like a buffer that quickly binds excess AS but rapidly releases the enzyme when its level decreases.

The SZ model equations were modified to simulate cases where some of the three negative-feedback inhibition mechanisms are lacking. Figure 4.3(c) shows both the case where AS inhibition is the only negative-feedback mechanism and the case where AS inhibition is absent. As shown in this figure, AS inhibition by itself (dashed line) immediately brings the system to a steady-state level of enzyme activity. If AS inhibition is absent (dot-dashed line), the AS activity can overshoot before reaching the steady state; this overshoot is explained by the time delay introduced by transcription and translation of the operon.

The influence of transcriptional attenuation is demonstrated in the simulation of a reactivation experiment shown in Fig. 4.3(d). A reactivation experiment starts with the initial conditions $M = E_{tot} = T_{tot} = 0$ (a derepression experiment has initial conditions $M = E_{tot} = 0$, but with initial T_{tot} being a positive constant). The model is transformed to one lacking the attenuation mechanism by making the probability associated with attenuation a constant fraction instead of the normal Trp-dependent expression of eqn 4.12. Figure 4.3(d) shows that the response without transcriptional attenuation (dashed curve) is slower than the normal case.

4.5 The lac operon

The lac operon is composed of genes coding for enzymes needed for processing lactose. The lac operon in *E. coli* has been studied in detail; its regulatory network is summarized in Fig. 4.4. As shown in this figure, the operon is composed of three structural genes – namely, *lacZ, lacY, lacA* – and the gene *lacI* that codes for a transcriptional repressor protein. Besides the promoter-operator region, a region called *cap* exists where a complex called CAP binds (CAP is made up of CRP and cAMP; CRP = cyclic AMP receptor protein, cAMP = cyclic adenosine monophosphate). The gene *lacZ* codes for the enzyme β-galactosidase and *lacY* codes for lactose permease, an enzyme that facilitates the flux of lactose through the cell membrane and into the cell. The gene *lacA* codes for thiogalactoside transacetylase involved in sugar metabolism but does not seem to play a direct role in the processing of lactose.

Glucose is the preferred carbon source of *E. coli*, but in the case where glucose is absent and lactose is present in the culture medium, the lac operon is switched on so that the external lactose is brought in and metabolized by the cell. The network in Fig. 4.4 explains the general observations summarized in Table 4.2 that shows that lactose induces the expression of lac operon genes (this is why lactose is often referred to as an *inducer*; note that allolactose is also often referred to as an *inducer*).

(a)

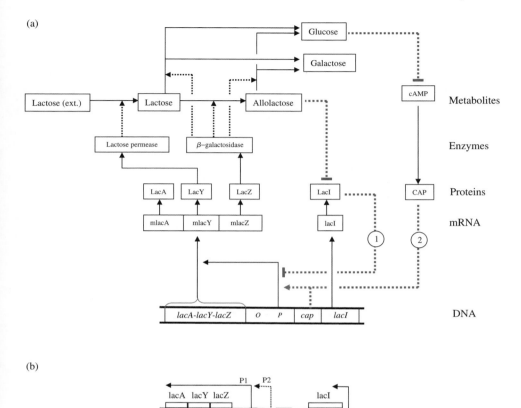

(b)

Fig. 4.4 (a) Regulation of expression of the lac operon. (1) = multimerization of lacI repressor molecules then binding to the operator-promoter region, (2) = CAP complex binding to cap region and enhances binding of RNA polymerase. (b) Details of the lac operon. The lacI repressor (tetramer) binds any or combinations of three operator regions O1, O2, and O3. RNA polymerase can bind to either promoter P1 or P2. The CAP complex binds the *cap* region.

Table 4.2 Turning on and off the expression of lac operon genes.

	(−) glucose	(+) glucose
(−) lactose	OFF	OFF
(+) lactose	ON	Low level of operon expression

If lactose is absent, the lac operon is OFF because the lacI repressor binds the operator-promoter region and prevents RNA polymerase from binding, regardless of glucose level (and despite possible CAP binding in the absence of glucose). If lactose and glucose are both present, expression of the operon is possible because the repressor is inhibited by allolactose, thereby allowing RNA polymerase to initiate transcription (but this expression occurs at low levels because CAP is prevented from binding the DNA). Only in the case where glucose is absent and lactose is present is the operon fully expressed.

4.6 Experimental evidence and modelling of bistable behavior of the lac operon

An elegant combination of modelling and experimental validation of the behavior of the lac operon in *E. coli* was performed by Ozbudak *et al.* (2004). The 'toy' model that these authors used to motivate their experiments focuses on the positive-feedback loop of the regulatory network of the operon (see Fig. 4.5).

The toy model involves two variables, x and y, whose dynamics are described by the following differential equations:

$$\tau_y \frac{dy}{dt} = \alpha \frac{1}{1 + (R/R_0)} - y, \tag{4.19}$$

$$\tau_x \frac{dx}{dt} = \beta y - x, \tag{4.20}$$

with R being a function of x according to

$$\frac{R}{R_T} = \frac{1}{1 + (x/x_0)^n}. \tag{4.21}$$

The meaning of the symbols are: R is the concentration of active lacI (the repressor), y is the concentration of the permease, and x is the intracellular concentration of the inducer; also, $n = 2$. R_T = total repressor concentration, and R_0 = initial repressor

Fig. 4.5 The toy model used by Ozbudak *et al.* (2004). x = inducer, R = active lacI (repressor), y = permease.

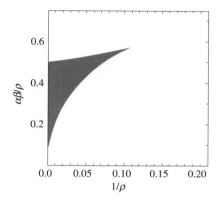

Fig. 4.6 Phase diagram of the model described by eqns 4.19 and 4.20. The parameter ρ is defined by: $\rho = 1 + (R_T/R_0)$. The grey region gives the set of parameters where 3 steady states coexist. Figure reproduced and modified from Ozbudak *et al.* (2004) with permission. Copyright 2004 Nature Publishing Group.

concentration. The steady states of the model are determined by equating the right-hand sides of eqns 4.19 and 4.20 to zero. Analysis of the steady states as functions of the parameters shows that there are parameter regions where only one steady state exists, and there are regions where three steady states can coexist, with the intermediate steady state being unstable. The phase diagram determined by Ozbudak *et al.* (2004) is shown in Fig. 4.6.

The experiments of Ozbudak *et al.* (2004) are summarized in Fig. 4.7. The phenomenon of bistability is well demonstrated in Figs. 4.7(b) and (c).

4.7 A reduced model derived from the detailed lac operon network

Santillan *et al.* (2007) modelled the bistability of the lac operon, including simulations of the experiments of Ozbudak *et al.* (2004). The discussion below provides important lessons in model reduction. The detailed network can be reduced to the following model involving three dynamical variables, namely, M = mRNA of *lacZ-lacY-lacA*, E = lacZ or lacY polypeptide, L = intracellular lactose:

$$\frac{dM}{dt} = k_M D P_D(G_e) P_R(A) - \gamma_M M$$

$$\text{(with } A = L \text{ as explained below)}, \tag{4.22}$$

$$\frac{dE}{dt} = k_E M - \gamma_E E, \tag{4.23}$$

$$\frac{dL}{dt} = k_L \beta_L(L_e)\beta_G(G_e)Q - 2\phi_M F(L)B - \gamma_L L$$

$$\text{(with } Q = E, \ B = E/4 \text{ as explained below)}. \tag{4.24}$$

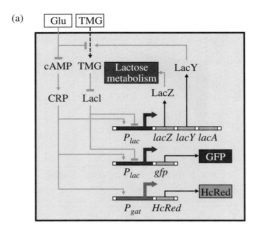

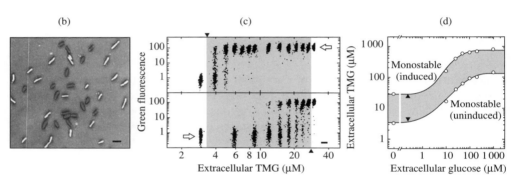

Fig. 4.7 (a) A simplified view of the lac operon regulatory network with the additional expression of fluorescent reporter proteins (GFP and HcRed) used in the experiments of Ozbudak *et al.* (2004). Instead of lactose, TMG (thio-methylgalactoside) is used as the inducer. TMG is not metabolized by β-galactoside and TMG uptake does not affect cAMP levels. Thus, TMG and glucose are used to independently regulate the activities of lacI and CRP. Transcriptional activity at the lac promoter (labelled P_{lac} in the figure) is reported by the green-fluorescent protein, GFP, while the effect of CRP on the gat (galactitol) promoter is reported by the red fluorescent protein HcRed (used to independently see the effect of CRP on the operon). (b) *E. coli* cells show coexistence between induced (*green*) and uninduced states; cells are grown in 18 μM TMG (which lies in the grey bistable region, see (c)). (c) Hysteresis experiments. The vertical axis (green fluorescence) corresponds to the number of cells that express GFP. *Lower panel* (uninduced to induced state transition): uninduced cells grown in increasing levels of TMG, starting with the value corresponding to the *white arrow*. *Upper panel* (induced to uninduced state transition): starting with a high level of TMG (*white arrow*), the value of TMG is progressively decreased. The grey region between 3 and 30 μM TMG is the bistable region. (d) Phase diagram showing the bistable, uninduced monostable, and induced monostable regions in the parameter space of extracellular glucose and extracellular TMG concentrations. Figure reproduced and modified from Ozbudak *et al.* (2004) with permission. Copyright 2004 Nature Publishing Group. (See Plate 2)

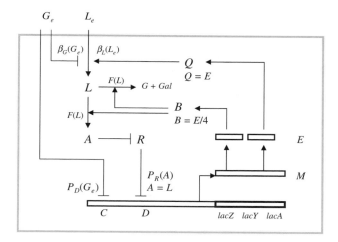

Fig. 4.8 The species labelled L, E, and M are the chosen dynamical variables of the Santillan *et al.* (2007) model. L = intracellular lactose, M = mRNA of *lacZ-lacY-lacA*, E = lacZ or lacY polypeptides, G_e = external glucose, L_e = external lactose, A = allolactose, R = repressor (lacI), B = β-galactosidase enzyme, Q = permease, $G + Gal$ = glucose and galactose, C = cap region on DNA, D = lac promoter. See text for the definitions of the functions $P_D(G_e)$, $P_R(A)$, $F(L)$, $\beta_G(G_e)$, and $\beta_L(L_e)$.

Figure 4.8 aids in understanding the above equations.

The first term on the right-hand side of eqn 4.22 incorporates the three main contributions to the production of the mRNA, namely, (1) the concentration D of lac promoter, (2) the probability $P_D(G_e)$ of having a RNA polymerase bound to the lac promoter taking into account concentrations of the CAP complex, RNA polymerase, and repression by external glucose, G_e (the expression for $P_D(G_e)$ is discussed below), and (3) the probability $P_R(A)$ that the lac promoter is *not* repressed as a function of inducer concentration A (the expression for $P_R(A)$ and the justification for equating the concentrations A and L are discussed below).

The second term on the right-hand side of eqn 4.22 represents the combined degradation and dilution (due to cell growth) of the mRNA. The rate coefficients k_M and γ_M are constant parameters of the model.

The first term on the right-hand side of eqn 4.23 is the rate of translation of the lacZ (also of lacY and lacA) polypeptide from the mRNA, and the second term gives the combined degradation and growth-dilution of the polypeptide. The rate coefficients of these processes are k_E and γ_E, respectively.

The first term on the right-hand side of eqn 4.24 gives the rate of increase of intracellular lactose as determined by the concentrations of external lactose (L_e), external glucose (G_e), and the lactose permease (Q). As indicated in Fig. 4.8, $Q = E$ since the permease is made up of a single lacY polypeptide (it is assumed that the levels of lacZ and lacY polypeptides are equal). The expressions for β_L and β_G are as follows:

$$\beta_L(L_e) = \frac{L_e}{\kappa_L + L_e}, \tag{4.25}$$

$$\beta_G(G_e) = 1 - \phi_G \frac{G_e}{\kappa_G + G_e}. \tag{4.26}$$

Thus, the maximum rate of increase of intracellular lactose is equal to $k_L Q$ that occurs when glucose is absent and $L_e \gg \kappa_L$.

The second term on the right-hand side of eqn 4.24 accounts for the two chemical processes (see Fig. 4.8) that consume intracellular lactose (hence the factor 2): conversion of lactose to allolactose, and hydrolysis of lactose to glucose and galactose; both chemical processes are catalyzed by β-galactosidase (B), and are assumed to have the same rate of $\phi_M F(L)B$, where

$$F(L) = \frac{L}{\kappa_M + L}. \tag{4.27}$$

Since β-galactosidase enzyme is made up of four subunits of the lacZ polypeptide, the concentration of the enzyme B is set equal to $E/4$ as indicated in Fig. 4.8. The rate coefficient ϕ_M and the Michaelis–Menten constant κ_M are constant parameters of the model. The last term on the right-hand side of eqn 4.24 is due to the growth-dilution and other processes that may consume lactose.

The probability of RNA polymerase binding, $P_D(G_e)$. Although RNA polymerase can bind promoter P1 (see Fig. 4.4(b)) on its own, the presence of a bound CAP on the *cap* region enhances the binding affinity of the polymerase. This co-operative binding can be shown (see Exercise 5) to lead to the following expression for the probability (P_D) of polymerase binding on the promoter as a function of polymerase concentration P and CAP concentration C:

$$P_D = \frac{\frac{P}{K_P}(1 + k_{pc}\frac{C}{K_C})}{1 + \frac{P}{K_P} + \frac{C}{K_C} + k_{pc}\frac{P}{K_P}\frac{C}{K_C}}, \tag{4.28}$$

where K_P is the equilibrium constant for the dissociation (reversible) of the polymerase–promoter complex, K_C is the equilibrium constant for the dissociation (reversible) of the CAP–*cap* complex, and k_{pc} (>1) is a constant that reflects the co-operation between the promoter and *cap* for enhanced binding of the RNA polymerase to the promoter. Note that eqn 4.28 can be rewritten as follows

$$P_D = \frac{p_P + p_P p_C(k_{pc} - 1)}{1 + p_P p_C(k_{pc} - 1)}, \tag{4.29}$$

where $p_P = (\frac{P}{K_P})/(1 + \frac{P}{K_P})$ is the probability of polymerase binding to the promoter without CAP, and $p_C = (\frac{C}{K_C})/(1 + \frac{C}{K_C})$ is the probability of CAP binding to *cap* without the polymerase. Now, note that the CAP complex concentration (C) is a function of the external glucose concentration (G_e) since it is known from experiments that increasing G_e leads to downregulation of cAMP and thereby decreasing C. The function p_C is then expressed in terms of G_e and assumed to follow a Hill-type functional form:

$$p_C(G_e) = \frac{K_G^{n_H}}{K_G^{n_H} + G_e^{n_H}}, \tag{4.30}$$

where n_H (the Hill exponent) and K_G are parameters of the model.

The probability that the promoter is not repressed, $P_R(A)$. As shown in Fig. 4.4(b) there are three operator regions, namely, O1, O2, and O3, in the lac operon. Binding of the repressor on O1 prevents initiation of transcriptions. Binding of a repressor to either O2 or O3 does not affect transcription, but DNA can fold in a way such that a single repressor molecule binds any two of these operators at the same time (for all combinations possible) thereby inhibiting initiation of transcription (see Santillan *et al.* 2007). It can be shown that the probability that the lac promoter is not repressed is

$$P_R = \frac{(1 + \frac{R}{K_2})(1 + \frac{R}{K_3})}{(1 + \frac{R}{K_1})(1 + \frac{R}{K_2})(1 + \frac{R}{K_3}) + R(\frac{1}{K_{12}} + \frac{1}{K_{13}} + \frac{1}{K_{23}})}, \tag{4.31}$$

where R is the concentration of the active repressor, K_i ($i =$ 1, 2, 3) the equilibrium constant of the reversible dissociation of the R–O$_i$ complex, and $K_{ij}(i, j = 1, 2, 3, i < j)$ is the equilibrium constant of the reversible dissociation of the O$_i$–R–O$_j$ complex. Considering that a repressor is a tetramer and that each subunit can be bound by an inducer molecule (identified as allolactose here), one can show that the concentration R of the active repressor is given by

$$R = R_T \left(\frac{K_A}{K_A + A} \right)^4, \tag{4.32}$$

where R_T is the total repressor concentration and K_A is the equilibrium constant for the allolactose–repressor subunit complex dissociation (reversible) reaction. Thus, the probability P_R in eqn 4.31 is a function of the inducer concentration A. $P_R(A)$ can be rewritten in the following form:

$$P_R(A) = \frac{1}{1 + \rho(A) + \frac{\varepsilon_{123}\rho(A)}{(1+\varepsilon_2\rho(A))(1+\varepsilon_3\rho(A))}} \tag{4.33}$$

where

$$\rho(A) = \rho_{\max} \left(\frac{K_A}{K_A + A} \right)^4, \quad \rho_{\max} = \frac{R_T}{K_1}$$

$$\varepsilon_i = \frac{K_1}{K_i} \ (i = 2, 3), \quad \text{and} \quad \varepsilon_{123} = \frac{K_1}{K_{12}} + \frac{K_1}{K_{13}} + \frac{K_1}{K_{23}}.$$

As indicated after eqn 4.22, A in $P_R(A)$ can be replaced by L (the lactose concentration); this can be justified if the rate of allolactose production from lactose and the

Table 4.3 Parameter values used in the model of Santillan *et al.* (2007). mpb = molecules per average bacterium.

$\mu = 0.02$ min^{-1}	$K_G = 2.6\,\mu$M
$D = 2$ mpb	$n_h = 1.3$
$k_M = 0.18$ min^{-1}	$\varepsilon_2 = 0.05$
$k_E = 18.8$ min^{-1}	$\varepsilon_3 = 0.01$
$k_L = 6.0 \times 10^4$ min^{-1}	$\varepsilon_{123} = 163$
$\gamma_M = 0.48$ min^{-1}	$\rho_{max} = 1.3$
$\gamma_E = 0.03$ min^{-1}	$K_A = 2.92 \times 10^6$ mpb
$\gamma_L = 0.02$ min^{-1}	$\kappa_L = 680\,\mu$M
$k_{pc} = 30$	$\phi_G = 0.35$
$p_p = 0.127$	$\kappa_G = 1.0\,\mu$M
$\phi_M = 0$ min^{-1} to 4×10^4 min^{-1}	$\kappa_M = 7 \times 10^5$ mpb

rate of allolactose metabolism to galactose and glucose are much faster than the cell growth rate.

Model simulations. With the parameter values given in Table 4.3, one can show that the system of model equations 4.22–4.24 possesses three steady states, the highest and lowest states being stable locally and the middle state being unstable. Thus, the model exhibits bistability.

Figure 4.9 gives the model's predicted phase diagrams on $G_e - L_e$ parameter space for increasing values of the parameter ϕ_M (the rate coefficient for lactose metabolism catalyzed by β-galactosidase). The parameters G_e and L_e – the extracellular glucose and lactose–are controlled in the experiments of Ozbudak *et al.* (2007) (recall that they used TMG instead of lactose). These phase diagrams show a connected region of bistability (grey region in Fig. 4.9), and monostable regions of uninduced and induced operon expression. 'Induced operon' means that steady-state expression of lacZ polypeptide (E in eqn 4.23) corresponds to the highest steady state; the 'uninduced operon' corresponds to the lowest steady state. Note that the experimental data points shown in Fig 4.7(d) are plotted in Fig. 4.9 (A) (the black dots) where agreement with the model prediction is very good. The Ozbudak *et al.* experiments can only be compared with Fig. 4.9 (A) where $\phi_M = 0$, because in the experiments the inducer TMG cannot be metabolized by β-galactosidase.

Ozbudak *et al.* (2004) reported that bistability was not observed when they repeated their experiments using lactose instead of TMG; this observation led to the question whether bistability is a natural property of the lac operon or an artifact of using non-metabolizable inducers. The numerical results of Santillan *et al.* (2007) shown in Fig 4.9(b)–(d) demonstrate that increasing ϕ_M increases the threshold levels of L_e for observing bistability. In addition, as ϕ_M increases, the bistable region shrinks – especially at low values of G_e as shown in Fig. 4.9(d). These may be the reasons, according to Santillan *et al.*, why bistability was not observed when lactose is used as the inducer.

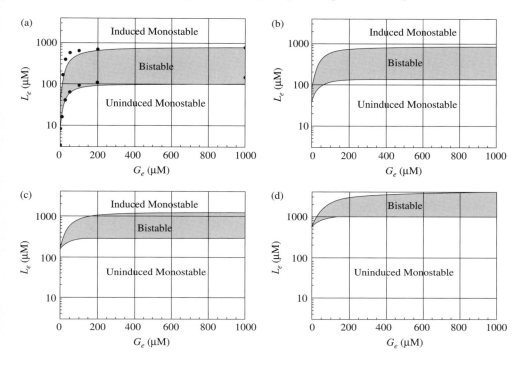

Fig. 4.9 Phase diagrams generated by the model of Santillan *et al.* (2007). L_e = extracellular lactose concentration, G_e = extracellular glucose concentration. Regions labelled 'uninduced monostable' and 'induced monostable' have unique low and high, respectively, steady states of the lacZ polypeptide. Region labelled 'bistable' means there are two coexisting stable steady states (with a third intermediate steady state that is unstable). The values of the parameter ϕ_M are varied as shown. Other parameter values are given in Table 4.3. Figures reproduced with permission from Santillan *et al.* (2007). Copyright 2007 Biophysical Society.

4.8 The challenge ahead: complexity of the global transcriptional network

The circular *E. coli* chromosome has approximately 4500 genes, and many of these genes belong to over 400 known operons. The discussion of the Trp and lac operons in this chapter has given a glimpse of the complexity of the transcriptional regulatory networks of this bacterium. A database accessible on the internet, *RegulonDB* (http://regulondb.ccg.unam.mx/index.html), provides information on the gene structure and the transcriptional regulation of *E. coli* operons. Figure 4.10 gives a picture of the regulatory interactions among 413 nodes (each node representing an operon) and 576 directed links (regulatory relationships between operons). *RegulonDB* also contains information on transcription factors (TFs) that regulate promoter activities of these operons. For the Trp operon, the repressor protein R is an example of a TF

Fig. 4.10 The transcriptional network of *E. coli* showing 413 nodes (each node representing an operon) and 576 directed links (regulatory relationships). The 10 *red circles* at the center (with labels) are 'global' regulators that have links to almost all of the peripheral operons. The operons are grouped (shown as clusters of circles, each cluster with a different color) into functional modules – modules with identifiable cellular functions, such as 'aromatic amino acid synthesis' (e.g. tryptophan), 'lactose transport and metabolism', etc. Figure reproduced with from Ma *et al.* (2004). Copyright 2004 Ma *et al.* (See Plate 3)

regulated by the metabolite tryptophan. For the lac operon, the lacI repressor and the CAP complex are TFs regulated by interactions with allolactose and glucose or cAMP, respectively. Thus, these TFs can be viewed as sensors for changes in metabolites in the extracellular or intracellular environment; changes that are then relayed to the DNA-based transcriptional machinery via the changes in TF binding to DNA. (For a more detailed discussion on this topic, the paper of Martinez-Antonio *et al.* (2007) listed in the References is recommended.) The challenge is to integrate these regulatory networks in ways that are amenable for further modelling and analysis. Groupings of the operons according to functional clusters or modules is a step in the right direction (see Fig. 4.10, and Ma *et al.* (2004)).

References

Lodish, H., Berk, A., Zipursky, S. L., Matsudaira, P., Baltimore, D., and Darnell, J. E. (1999) *Molecular cell biology*. W. H. Freeman & Co., New York.

Ma, H-W., Buer, J., and Zeng, A-P. (2004) 'Hierarchical structure and modules in the *Escherichia coli* transcriptional regulatory network revealed by a new top-down approach,' *BMC Bioinformatics* **5**, 199.

Martinez-Antonio, A., Janga, S. C., Salgado, H., and Collado-Vides, J. (2006) 'Internal-sensing machinery directs the activity of the regulatory network in *Escherichia coli*,' *Trends in Microbiology* **14**, 22–27.

Ozbudak, E. M., Thattai, M., Lim, H. N., Shraiman, B. I., and van Oudenaarden, A. (2004) 'Multistability in the lactose utilization network of *Escherichia coli*,' *Nature* **427**, 737–740.

Santillan, M., Mackey, M. C., and Zeron, E. S. (2007) 'Origin of Bistability in the *lac* Operon,' *Biophysical Journal* **92**, 3830–3842.

Santillan, M. and Zeron, E. S. (2004) 'Dynamic influence of feedback enzyme inhibition and transcription attenuation on the tryptophan operon response to nutritional shifts,' *Journal of Theoretical Biology* **231**, 287–298.

Exercises

1. Derive eqns 4.7 and 4.10.
2. Show that the factor $1/2$ in eqn 4.2 occurs because two TrpE proteins are required to assemble one AS molecule, and with the assumption that the dissociation rate is small. *Hint*: Let A be a monomer and B a dimer. Consider the reversible reaction $2A \overset{k_+, \ k_-}{\longleftrightarrow} B$, where k_+ and k_- are the rate coefficients of the forward and reverse reactions, respectively. At chemical equilibrium, $[A]_{eq}^2 = K_D[B]_{eq}$ where $K_D = k_-/k_+$ and the subscript *eq* refers to equilibrium concentration. Let $[A]_T = [A] + 2[B]$ be the total monomer concentration. Show that $[B]_{eq} \sim \frac{1}{2}[A]_{eq}$ if $K_D << [A]_T$.
3. Verify numerically the phase diagram in Fig. 4.6.
4. Consider a molecule A with specific *independent* binding sites for N different molecules B_i ($1 \leq i \leq N$). Let $(n_1, \ldots, n_N) \equiv \{n_i\}$ denote the bound state of A, with $n_i = 1$ if a molecule B_i is bound to its specific site, and $n_i = 0$ otherwise. Prove that the probability of a specific bound state $(n_1, \ldots, n_N)$ is given by

$$P_N(n_1, \ldots, n_N) = \frac{\prod_{i=1}^{N} ([B_i]/K_i)^{n_i}}{\sum_{\{n_j\}} \prod_{j=1}^{N} ([B_j]/K_j)^{n_j}},$$

where K_i is the equilibrium (dissociation) constant of the AB_i complex, and the sum over all $\{n_j\}$ in the denominator means over all the possible 2^N bound states of A. *Hint*: For the case $N = 1$, one has $[B_1][A]_{eq} = K_1[AB_1]_{eq}$ at equilibrium. Using $[A]_{total} = [A] + [AB_1]$, the following equations hold: $\frac{[A]}{[A]_{total}} = \frac{1}{1 + \frac{[B_1]}{K_1}} \equiv P(n_1 = 0)$, and $\frac{[AB_1]}{[A]_{total}} = \frac{\frac{[B_1]}{K_1}}{1 + \frac{[B_1]}{K_1}} \equiv P(n_1 = 1)$. Hence, $P(n_1) = \frac{\left(\frac{[B_1]}{K_1}\right)^{n_1}}{1 + \frac{[B_1]}{K_1}}$. Now consider $N = 2$

and use the information that binding to all sites is independent of each other; and finally, by induction, arrive at the desired general formula for $P_N(n_1,\ldots,n_N)$.

5. If in Exercise 3 two sites, p and q, have a co-operative interaction in the sense that the probability of finding both B_p and B_q bound to A is stronger by a factor k_c (>1) than the corresponding probability if the two binding sites were independent, then

$$
P(n_1,\ldots,n_N) = \frac{k_c^{n_p n_q} \prod\limits_{i=1}^{N} ([B_i]/K_i)^{n_i}}{\sum\limits_{\{n_j\}} k_c^{n_p n_q} \prod\limits_{j=1}^{N} ([B_j]/K_j)^{n_j}}.
$$

Use this formula to verify eqn 4.28.

6. Verify eqn 4.31.

5
Control of DNA replication in a prokaryote

A cell-division cycle, or *cell cycle* for short, is a combination of growth to double the amounts of cellular components, and division of these components between two daughter cells. The chromosomes must be faithfully replicated once during the cell cycle and segregated equally so that each daughter cell gets a full complement of genes. The cell cycle of *Escherichia coli*, a common bacterium that thrives in the guts of humans and other animals, is the subject of this chapter. The focus of computational models of *E. coli* replication is the co-ordination between initiation of DNA replication and cell growth so that the average size of the bacterium is maintained constant throughout generations.

5.1   The cell cycle of *E. coli*

E. coli is a rod-shaped bacterium with a diameter in the range of 0.3–0.7 μm and length in the range of 0.5–3 μm (see Fig. 5.1). Its cell-cycle period, commonly referred to as the *generation time*, can be as short as 20 min (as in optimum conditions at 37 °C) or as long as several hours (as in minimal media where *E. coli* synthesizes molecular building blocks from simple carbon sources). For slow-growing cultures (those with generation times longer than 80 min), the cell cycle resembles that of eukaryotes. The cell-cycle phases are labelled with the letters B, C, and D – roughly corresponding to the G1, S, and G2/M phases of eukaryotic cell cycles (discussed in Chapters 6 and 7). An example of an *E. coli* cell cycle is shown in Fig. 5.2. Note that unlike in eukaryotes where DNA replication can initiate at multiple regions in the chromosome, the initiation of replication in *E. coli* occurs at a single unique site on the DNA; this well-characterized site is called the oriC and is described in Section 5.3. In Fig. 5.2, the single *E. coli* circular chromosome is depicted as a circle with a green dot representing the oriC and the yellow dot indicating the location of the replication complex that moves along the DNA during replication. The replication complex is a molecular machine whose components include DNA polymerases (enzymes that copy DNA) and DNA helicases (enzymes that unzip double-stranded DNA).

A key feature of bacterial cell cycles is the requirement for a minimum mass to initiate chromosome replication. This initiation mass is attained at the end of the B-period. The chromosome is replicated during the C-period; and the D-period is used for segregating the chromosomes between the two daughter cells. The average cell

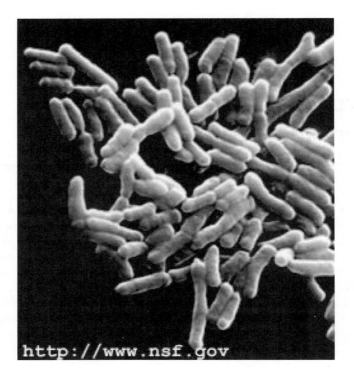

Fig. 5.1 A colony of the rod-shaped bacterium *Escherichia coli*. Picture courtesy of the National Science Foundation (USA).

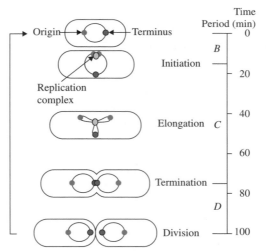

$\tau = 100$ min, $B = 15$ min, $C = 60$ min, $D = 25$ min

Fig. 5.2 The bacterial cell cycle in slow-growing cultures. The cell-cycle period is also called the generation time τ. Figure reproduced with permission from T. Atlung (2004). (See Plate 4)

mass, M, increases exponentially throughout the cell cycle:

$$M(t) = M(0)\exp(\mu t) \quad \text{for } 0 \leq t \leq \tau, \tag{5.1}$$

where μ is a growth-rate constant, and τ is the generation time equal to the sum of the periods $(B + C + D)$. Since mass doubles in a generation, i.e. $M(\tau) = 2M(0)$, one obtains

$$\tau = \frac{\ln 2}{\mu}. \tag{5.2}$$

There are two periodic processes in a cell cycle. One is a growth process that begins with one cell and ends with two cells (this cycle's period is τ); the other is a chromosome-replication cycle that begins with one circular chromosome and ends with two chromosomes (this cycle's period is C). The cell must co-ordinate these two cycles in order to proliferate successfully. This co-ordination problem is discussed next.

5.2 Overlapping cell cycles: coordinating growth and DNA replication

For a wide range of growth rates or generation times, it is known that the periods C and D are approximately constant. C is about 40 min and D is about 20 min. Thus, for generation times greater than 60 min, period B increases correspondingly. Cases of generation times less than $(C + D)$ do exist – these are cases where over-lapping cell cycles occur. An example is shown in Fig. 5.3 where the generation time is 50 min.

Figure 5.3 shows the solution for the problem of co-ordinating a growth cycle with period $\tau = 50$ min and chromosome cycle with period $C = 40$ min under the constraint that D must be equal to 20 min. The C period starts at $t = 40$ min (labelled 'Initiation') in one cell cycle and ends at $t = 30$ min (labelled 'Termination') in the *next* cell cycle. The D period (20 min) starts at $t = 30$ min (labelled 'Termination') in a cell cycle and ends at $t = 50$ min (labelled 'Division') in the same cell cycle. Thus, as shown at $t = 0$ min in Fig. 5.3, each newborn cell contains a replicating chromosome with two origins (depicted by green dots). In one of the exercises at the end of this chapter, the case of a bacterial cell with 25 min generation time is considered. In this case, each newborn cell has to initiate at four origins of replication.

5.3 The oriC and the initiation of DNA replication

Figure 5.4(a) shows a micrograph of the *E. coli* chromosome attached to some sub-stratum. This single-circular chromosome is composed of (among other things) a DNA duplex with 4.64 million nucleotide base pairs containing approximately 4500 genes. The chromosome is nearly 1 mm long when stretched, but it occupies only a third to a half of the cell volume ($\sim 1\,\mu\text{m}^3$) in its supercoiled state in the cell. Figure 5.4(b) is a picture of an *E. coli* chromosome that is more than half-way to finishing replication. The advancing replication fork is indicated in the inset at the top right corner (note that these two replication forks are very close to each other in a cell – this is why

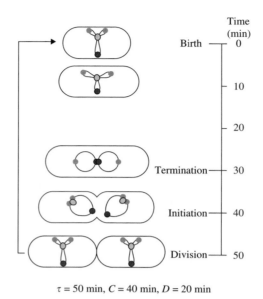

$\tau = 50$ min, $C = 40$ min, $D = 20$ min

Fig. 5.3 The bacterial cell cycle in fast-growing cultures. Figure reproduced with permission from T. Atlung (2004). (See Plate 5)

they are shown as single yellow dots in Fig. 5.2 and Fig. 5.3). Backward in time, these two replication forks would have originated from a region in the DNA called the *oriC*. Chromosomal replication in *E. coli* always initiates at a unique oriC region whose genomics (that is, the DNA base sequence) has been well characterized.

The oriC of *E. coli* has five sites called *9-mers* after a consensus 9-base sequence TTAT(C/A)CA(C/A)A, where the (C/A) means either C or A. These sites are also called *dnaA boxes* because they are the sites where the *dnaA* protein binds the DNA to initiate replication. As will be discussed later, the binding of dnaA proteins on the oriC is key to understanding the control of initiation of DNA replication. Outside the oriC, there are about 300 dnaA boxes distributed throughout the chromosome to which dnaA can also bind; however, even if these non-oriC boxes are bound by dnaA proteins, replication does not initiate on any of them – this is why they are referred to as non-functional boxes.

The dnaA protein in *E. coli* exists either in an ATP-bound or an ADP-bound form. *In-vitro* studies have shown that only dnaA-ATP is able to initiate replication at an oriC, although both forms are able to bind all the dnaA boxes. Twenty to forty dnaA-ATP monomers bind the oriC co-operatively, and it is believed that binding of a minimum number of dnaA-ATP monomers is required to unwind the DNA and initiate replication. This unwinding is followed by the unzipping (opening) of a segment of the DNA duplex that allows the assembly of multiprotein complexes that copy the DNA strands.

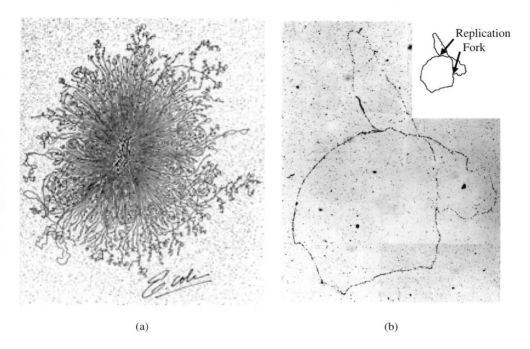

(a)   (b)

Fig. 5.4 (a) Micrograph of an *E. coli* chromosome. (b) Autoradiograph of a replicating *E. coli* chromosome. Picture reproduced with permission. Copyright Prentice-Hall, Inc.

Another important feature of the *E. coli* cell cycle is the fact that dnaA-ATP cannot bind to the newly replicated oriCs immediately. There is a refractory period – the so-called *eclipse period* – of 8 to 10 min before dnaA can bind again to the oriC boxes. Although not fully understood at this time, the eclipse period could be due to the methylation state of the DNA after replication, and also to the observed binding of the newly replicated oriCs to the cell membrane. Furthermore, upon initiation of replication the oriC-bound dnaA-ATP is converted to the dnaA-ADP form that is unable to initiate replication (dnaA-ADP is assumed to fall off the oriC).

5.4 The initiation-titration-activation model of replication initiation

One of the earliest models of DNA replication initiation in bacteria is due to Pritchard and coworkers (1969). The model described in this section is called the *initiation-titration-activation model* due to Browning *et al.* (2004); the model incorporates the earlier models of Mahaffy and Zyskind (1989), and Hansen *et al.* (1991). In the following discussion, dnaA boxes within and outside the oriC regions will be often referred to simply as *oriC boxes* and *non-oriC boxes*, respectively.

Two types of concentrations will be considered in the rate expressions included in the model. One refers to the number of monomers *per unit volume* of the cell's cytoplasm, and the other refers to the number of bound monomers *per dnaA box* on the chromosome. Due to the small numbers of molecules involved, the proper way of modelling the dynamics of the system is to employ stochastic and spatially dependent simulation methods. However, to keep the discussion simple for now, it will be assumed that the time evolution of *average* concentrations can be described by deterministic differential equations.

The steps involved in the initiation of chromosome replication are depicted in Fig. 5.5.

5.4.1   DnaA protein synthesis

As shown in Fig. 5.5, the *dnaA* gene is located near the oriC. Expression of this gene has been shown to be inhibited by its product, the dnaA protein (autorepression). More specifically, dnaA-ATP and dnaA-ADP can bind to a region in the *dnaA* gene's promoter and repress transcription. DnaA rapidly binds ATP, so it is assumed that *dnaA* gene expression gives rise immediately to dnaA-ATP as soon as the protein is synthesized (note that ATP is abundant in the cell). The rate, v_1, of *dnaA* gene expression is thus assumed to be identical to the rate of formation of cytoplasmic (free) dnaA-ATP, and that this rate is proportional to the growth rate of the cell's mass and repressed by bound dnaA proteins:

$$v_1 = k_1 \left(\frac{\mathrm{d}M}{\mathrm{d}t} \right) \left(\frac{1}{1 + \alpha A_g + \beta A_{i,g}} \right), \tag{5.3}$$

where $\mathrm{d}M/\mathrm{d}t$ (equal to μM) is the growth rate of the cell mass, A_g is the number of dnaA-ATP monomers bound per non-oriC box, and $A_{i,g}$ is the number of dnaA-ADP

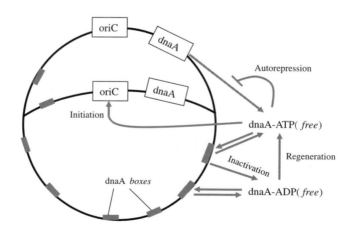

Fig. 5.5 A schematic diagram showing the steps in the initiation-titration-activation model of DNA replication initiation. Figure adapted from Browning *et al.* (2004).

monomers bound per non-oriC box. Strictly speaking, one should only consider in the last factor (for autorepression) in eqn 5.3 the dnaA proteins bound to oriC boxes since these are the proteins that directly repress the *dnaA* gene's transcription, but since the number of oriC boxes is very small compared to non-oriC boxes, the chance of binding to oriC boxes increases only after most of the non-oriC boxes are bound – this is what is meant by *titration* of dnaA proteins by the non-oriC boxes. The extent of autorepression is varied using the parameters α and β (Browning *et al.* (2004) use the following arbitrary values: $\alpha = 2, \beta = 0.02$). The rate coefficient k_1 subsumes a factor that considers the fraction of the contribution of dnaA synthesis to the overall growth rate of cell mass; this factor was explicitly stated in the model of Browning *et al.* (2004) where the fraction is proportional to the ratio of *dnaA* gene dosage to the total gene dosage:

$$k_1 = k_1' \left(\frac{dnaA \text{ gene dosage}}{\text{total gene dosage}} \right). \tag{5.4}$$

Browning *et al.* (2004) used the values $k_1' = 1.08 \times 10^{18}$ (to fit with the observed protein synthesis at 1000 to 2000 monomers per cell) and 0.018 for the gene-dosage ratio. DnaA is a stable protein and its degradation has been ignored in the model.

A key feature of the *E. coli* cell cycle is that the cell mass at which replication initiates (the initiation mass) is roughly constant for a wide range of growth rates. Thus, eqn 5.3 provides a link between this observation and the amount of dnaA proteins synthesized by the cell.

5.4.2 DnaA binding to boxes and initiation of replication

DnaA-ATP monomers bind to oriC boxes and non-oriC boxes as represented by the following steps:

$$A_f + n_o \xrightleftharpoons{2} A_o, \tag{5.5}$$

$$A_f + n_g \xrightleftharpoons{3} A_g, \tag{5.6}$$

where A_f is the symbol for free (cytoplasmic) dnaA-ATP, A_o is for dnaA-ATP bound to an oriC box, and A_g is for dnaA-ATP bound to a non-oriC box. The binding rates are proportional to the number of free oriC and free non-oriC boxes symbolized by n_o and n_g, respectively. The double-headed arrows in eqns 5.5–5.6 indicate reversible steps. Browning *et al.* (2004) took into account the different classes of dnaA boxes with different affinities to dnaA proteins (that is, different binding and dissociation rate constants for dnaA-ATP and dna-ADP); to simplify the discussion, only one class of dnaA boxes is considered here. Using the same monomer symbols for their respective concentrations, the two processes in eqns 5.5 and 5.6 occur at the following rates:

$$v_2 = k_2 n_o \left(\frac{A_f^c}{k_b + A_f^c} \right) \chi(t), \tag{5.7}$$

$$v_3 = k_3 n_g A_f, \tag{5.8}$$

where k_2 and k_3 are rate constants of the forward steps in eqns 5.5 and 5.6, respectively. The co-operative binding of dnaA-ATP monomers on the oriC boxes is represented by the Hill-type function $(A_f^c/(k_b + A_f^c))$ with exponent c (greater than 1) in eqn 5.7. The factor $\chi(t)$ in 5.7, accounting for the eclipse period (from the initiation time t^i to the end of the eclipse period t^e), is defined as

$$\chi(t^i \le t \le t^e) = 0 \quad \text{and} \quad \chi(0 < t < t^i) = \chi(t^e < t \le \tau) = 1. \tag{5.9}$$

The observed duration of the eclipse period (that is, $t^e - t^i$) is typically 8 to 10 min in *E. coli*. During the time interval between t^i and t^e, dnaA-ATP and dnaA-ADP cannot bind to oriC boxes; in contrast, binding of dnaA-ATP to non-oriC boxes (see eqn 5.6) is assumed to occur throughout the cell cycle. Initiation is triggered when A_o reaches a critical value, A^i, which is known to correspond to about 30 monomers of dnaA-ATP bound to the total of 5 oriC boxes.

The binding of dnaA-ATP to the boxes are reversible, with the following dissociation rates:

$$v_{2r} = k_{2r} A_o, \tag{5.10}$$

$$v_{3r} = k_{3r} A_g, \tag{5.11}$$

where k_{2r} and k_{3r} are rate constants of the reverse directions of eqns 5.5 and 5.6, respectively. Browning *et al.* (2004) cited three different values of dissociation rate constants depending on the affinity class of the dnaA boxes (ranging from 0.79 nM to 2×10^6 nM). Again, to simplify the discussion, only one affinity class is considered here.

Similar to the binding of dnaA-ATP to the boxes, dnaA-ADP can also reversibly bind to the boxes according to the following processes and rates:

$$A_{i,f} + n_o \overset{4}{\longleftrightarrow} A_{i,o} \tag{5.12}$$

$$A_{i,f} + n_g \overset{5}{\longleftrightarrow} A_{i,g}, \tag{5.13}$$

$$v_4 = k_4 n_o A_{i,f} \chi(t), \tag{5.14}$$

$$v_{4r} = k_{4r} A_{i,o}, \tag{5.15}$$

$$v_5 = k_5 n_g A_{i,f}, \tag{5.16}$$

$$v_{5r} = k_{5r} A_{i,g}, \tag{5.17}$$

where $\chi(t)$ is the eclipse factor given in eqn 5.9, $A_{i,f}$ is for cytoplasmic (free) dnaA-ADP, $A_{i,o}$ is for dnaA-ADP bound to oriC boxes, and $A_{i,g}$ is for dnaA-ADP bound to non-oriC boxes. Note that dnaA-ADP competes against dnaA-ATP for the available boxes n_o and n_g; expressions for n_o and n_g are discussed next.

5.4.3 Changing numbers of oriCs and dnaA boxes during chromosome replication

The numbers of free boxes, n_o and n_g, change with time because of the increasing chromosome length during replication. For simplicity, it is assumed that dnaA boxes are distributed uniformly around the chromosome so that the extent, E, of the chromosome can be measured in terms of the number of boxes. Let $E_m(t)$ be the length (in units of dnaA boxes) of the *additional* chromosome that is synthesized after initiation at the mth oriC (where m is an arbitrary indexing label given to existing oriCs; an oriC ceases to exist once it is replicated, giving two new ones). For a single chromosome, let there be ζ_o oriC boxes and ζ_g non-oriC boxes (in *E. coli*, $\zeta_o = 5$ and $\zeta_g \sim 300$). The maximum number of free dnaA boxes in one non-replicating chromosome with one oriC is then $\zeta = (\zeta_o + \zeta_g) \sim 305$. Assuming that the rate of replication is constant for all initiated oriCs, the rate of increase in chromosome length due to initiation at the mth oriC is

$$\frac{dE_m}{dt} = \frac{\zeta}{C} \quad \text{for } t_m^i \leq t \leq (t_m^i + C), \tag{5.18}$$

where C is the length of the replication period (the C-period) and t_m^i is the initiation time at the mth oriC. For example, for $C = 40$ min and $\zeta = 305$, the number of additional dnaA boxes created per minute during replication from one initiated oriC is about 7.6. Integrating the equation above, one obtains

$$E_m(t) = E_m(t_m^i) + (\zeta/C)t \quad \text{or}$$
$$E_m(t) = (\zeta/C)t \quad \text{for } t_m^i \leq t \leq (t_m^i + C), \tag{5.19}$$

since $E_m(t_m^i)$ is zero. The total length of the chromosome would then be

$$E_{\text{tot}}(t) = E_o + \sum_{m=1}^{q} E_m(t), \tag{5.20}$$

where the constant E_o is the starting length prior to initiation at the first oriC (i.e. $E_0 = \zeta = 305$) and q is the number of oriCs that initiate during a cell cycle. Thus, the number of free oriC and free non-oriC boxes vary with time as follows:

$$n_o(t) = (\zeta_o/\zeta)E_{\text{tot}}(t)\left(1 - \frac{A_o + A_{i,o}}{A*}\right), \tag{5.21}$$

$$n_g(t) = (\zeta_g/\zeta)E_{\text{tot}}(t)\left(1 - \frac{A_g + A_{i,g}}{A*}\right), \tag{5.22}$$

where $A*$ is the maximum number of bound monomers per box. An assumption made here is that both oriC and non-oriC boxes have identical maximum number capacities, each equal to $A*$ monomers per box where $A* = A^i \sim 6$.

5.4.4 Death and birth of oriCs

Once replication initiates at the mth oriC, this oriC disappears and gets replaced by two new ones. The two new oriCs are initialized with $A_o = 0$ and $A_{i,o} = 0$; this initialization implies that the dnaA-ATP and dnaA-ADP monomers bound to the oriC at the time of initiation are all dislodged back to the cytoplasm (with the dnaA-ATP converted to dnaA-ADP), contributing to the jump in $A_{i,f}$ after initiation. In general agreement with observations, it is assumed that the dynamics of sister oriCs (those coming from one oriC) are perfectly synchronized.

5.4.5 Inactivation of dnaA-ATP

As the replication fork passes by dnaA boxes bound by dnaA-ATP, ATP is hydrolyzed to ADP to generate the inactive dnaA-ADP form. Only inactivation at the non-oriC boxes will be considered in the model because, immediately after initiation, all monomers bound at the oriC boxes are dislodged and returned to their free forms in the cytoplasm. The inactivation process and its rate are given below:

$$A_g \xrightarrow{6} A_{i,g}, \qquad (5.23)$$

$$v_6 = k_6(\mathrm{d}E_{\mathrm{tot}}/\mathrm{d}t)A_g, \qquad (5.24)$$

where E_{tot} is given by Eqn. 5.20. Finally, in the cytoplasm, the free dnaA-ADP and free dnaA-ATP can be transformed to each other:

$$A_{i,f} \xleftrightarrow{7} A_f, \qquad (5.25)$$

$$v_7 = k_7 A_{i,f}, \qquad (5.26)$$

$$v_{7r} = k_{7r} A_f, \qquad (5.27)$$

where k_7 and k_{7r} are rate constants.

5.5 Model dynamics

The dynamics of chromosome replication coupled with cell growth is modelled by the differential eqns 5.28–5.33 given below. The expressions for the rates v_i have been described in the preceding section. The terms $(-\mu A_f)$ and $(-\mu A_{i,f})$ found on the right-hand sides of eqns 5.28 and 5.31, respectively, represent rates of decrease in cytoplasmic concentrations due to dilution as the cell volume grows exponentially.

The case of slow-growing *E. coli* cells (see Fig. 5.2) will be used to illustrate how a computer simulation is performed. At $t = 0$, the system starts with one oriC ($m = q = 1$) and with initial values for $A_f, A_{i,f}, A_o, A_{i,o}, A_g, A_{i,g}$, and cell mass M. For the entire generation time ($t = 0$ to $t = \tau$), an individual cell's mass increases exponentially according to eqn 5.1. The following differential equations are first integrated from $t = 0$ until the oriC boxes achieve the threshold value of A^i (dnaA-ATP monomers per oriC box) in order to determine the initiation time, t_m^i,

$$\frac{\mathrm{d}A_f}{\mathrm{d}t} = v_1 + (-v_2 + v_{2r}) + (-v_3 + v_{3r}) + (v_7 - v_{7r}) - \mu A_f, \tag{5.28}$$

$$\frac{\mathrm{d}A_o}{\mathrm{d}t} = v_2 - v_{2r}, \tag{5.29}$$

$$\frac{\mathrm{d}A_g}{\mathrm{d}t} = (v_3 - v_{3r}) - v_6, \tag{5.30}$$

$$\frac{\mathrm{d}A_{i,f}}{\mathrm{d}t} = (-v_4 + v_{4r}) + (-v_5 + v_{5r}) + (-v_7 + v_{7r}) - \mu A_{i,f}, \tag{5.31}$$

$$\frac{\mathrm{d}A_{i,o}}{\mathrm{d}t} = v_4 - v_{4r}, \tag{5.32}$$

$$\frac{\mathrm{d}A_{i,g}}{\mathrm{d}t} = (v_5 - v_{5r}) + v_6. \tag{5.33}$$

Immediately after initiation, the oriC ($m = 1$) disappears and gets replaced by two sister oriCs with indices $m = 2$ and $m = 3$. These two new oriCs are assumed to be perfectly synchronized (that is, they initiate and replicate at the same rate). The increasing length of the chromosome (due to the initiation at the first oriC) is given by eqn 5.20 (with $q = 1$). At $t = t_m^i$, the cytoplasmic monomer concentrations are increased (discontinuously) by the following amounts due to the monomers dislodged from the initiated oriC:

$$\Delta A_f = \zeta_o A_i / V(t_m^i), \tag{5.34}$$

$$\Delta A_{i,f} = \zeta_o A_{i,o}(t_m^i) / V(t_m^i), \tag{5.35}$$

where $V(t_m^i)$ is the volume of the cell at $t = t_m^i$. Note that $V(t_m^i) = V(0) \exp(\mu t_m^i)$ with the assumption that the cell (cytoplasm) density is kept constant as the cell grows. Integration of eqns 5.28–5.33 continues until $t = \tau$ when the mass $M(\tau)$ and the contents of the cell are halved prior to the start of the next cell cycle. Care must be taken to observe the eclipse period – see eqns 5.7–5.9 and 5.14 – and the end of the chromosome replication at $t = C$ (eqns 5.18 and 5.19).

5.6 Robustness of initiation control

The kinetic model described in the preceding section is identical in essence to a more complex deterministic model due to Browning *et al.* (2004); their model includes three different types of non-oriC dnaA boxes according to the binding affinities and dissociation rates of dnaA-ATP and dnaA-ADP monomers (high, medium, and non-specific affinities). The computer simulation shown in Fig. 5.6 demonstrates that the total dnaA protein level in a cell does not fluctuate much in comparison with the level of dnaA-ATP. Large-amplitude fluctuations of the replication initiator (that is, dnaA-ATP) alleviate the effects of noise that is necessary for robust control of replication initiation. The investigation of the robustness of the initiation control is described below.

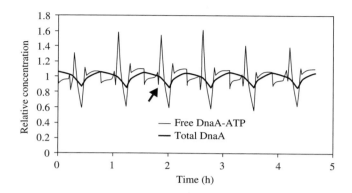

Fig. 5.6 Dynamics in an *E.*coli cell of the total dnaA protein concentration and the free dnaA-ATP concentration from a computer simulation using the model of Browning *et al.* (2004). The *arrow* indicates the time when DNA replication initiates. Reproduced with permission from Browning *et al.* (2004). Copyright 2004 Wiley Periodicals, Inc.

Browning *et al.* (2004) formulated a hybrid model that combines deterministic and stochastic steps in their investigation of the factors affecting the control of initiation. Steps 5.5, 5.6, 5.12, 5.13, and 5.23 are considered to occur stochastically because of the small numbers of dnaA boxes (n_o and n_g); the dynamics of the rest of the model is described by deterministic differential equations. The Gillespie algorithm (see Chapter 3) is used to determine the dynamics of the stochastic steps. The robust control of initiation is manifested in the almost constant interinitiation time, δt^i, that is, the time between two successive initiations. The doubling time of an *E. coli* population is defined as the average $\langle \delta t^i \rangle$, and a low standard deviation of interinitiation time, σ_{init}, is used as the measure of the robustness of the control mechanism of initiation. Figures 5.7 (a)–(d) summarize the results.

For rates of dnaA protein synthesis to the left of the vertical dotted line shown in Fig. 5.7(a), not enough dnaA is made to keep up with the growth rate, so that initiation may not occur at all. (The stepwise drops in σ_{init} on the right of Fig. 5.7(a) occur when the number of oriCs doubles at initiation.) Figure 5.7(b) shows that as long as the binding-rate constant is greater than ~0.1 $\text{M}^{-1}\text{s}^{-1}$, the standard deviation σ_{init} is insensitive to the rate of binding of dnaA-ATP monomers to oriC boxes. The dotted vertical line shown in the figure indicates the estimated experimental value of the binding-rate constant. Similarly, σ_{init} is generally insensitive to the binding constants for non-functional (non-oriC) boxes except when the binding rates are comparable to those of the binding rates at the oriC boxes – as demonstrated by the peak in σ_{init} in the middle of Fig. 5.7(c); in this case, binding of the initiator dnaA-ATP to non-oriC boxes interfere with binding to the oriCs. The independence of σ_{init} to low non-oriC binding constants – as shown on the left side of Fig. 5.7(c) – is not robust with respect to changes in the number of non-oriC boxes, as illustrated by the upper curve of Fig. 5.7(d). In contrast, when the non-oriC binding constant is high and when

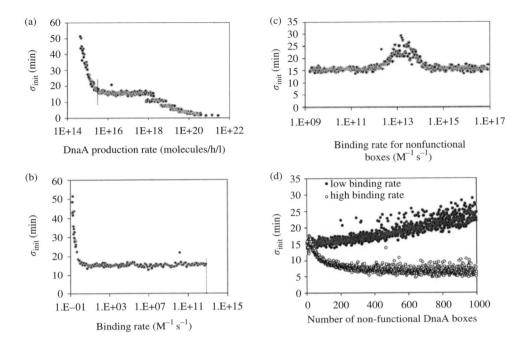

Fig. 5.7 Standard deviations of interinitiation times (σ_{init}) as functions of (a) rate of dnaA protein synthesis, (b) binding rate constants of dnaA-ATP to oriC boxes, (c) binding rate constants of dnaA-ATP to non-oriC (nonfunctional) boxes, and (d) number of non-functional dnaA boxes. The *dotted vertical line* in (a) indicates the observed experimental value of 3.5×10^{15} molecules/h per liter. The *dotted vertical line* in (b) shows the estimated experimental value of 9×10^{12} $\text{M}^{-1}\text{s}^{-1}$. Reproduced with permission from Browning *et al.* (2004). Copyright 2004 Wiley Periodicals, Inc.

the number of non-oriC boxes is large enough (>300), the interinitiation time becomes robust against further increases in the numbers of non-functional boxes.

References

Browning, S. T., Castellanos, M., and Shuler, M. L. (2004) 'Robust control of initiation of prokaryotic chromosome replication: Essential considerations for a minimal cell,' *Biotechnology and Bioengineering* **88**, 575–584.

Hansen, F. G., Christensen, B. B., and Atlung, T. (1991) 'The initiator titration model: computer simulation of chromosome and minichromosome control,' *Research in Microbiology* **142**, 161–167.

Mahaffy, J. M. and Zyskind, J. W. (1989) 'A model for the initiation of replication in *Escherichia coli*,' *Journal of Theoretical Biology* **140**, 453–477.

Pritchard, R. H., Barth, P. T., and Collins, J. (1969) 'Control of DNA synthesis in bacteria,' *Symposium of the Society for General Microbiology* **19**, 263–297.

Exercises

1. The terms $(-\mu A_f)$ and $(-\mu A_{i,f})$ found on the right-hand sides of $\mathrm{d}A_f/\mathrm{d}t$ and $\mathrm{d}A_{i,f}/\mathrm{d}t$ – eqns 5.28 and 5.31, respectively – represent rates of decrease in cytoplasmic concentrations due to dilution as the cell grows. In this case, the density of the cell is assumed constant as the cell grows. Show that if the cell mass increases exponentially, $\mathrm{d}M/\mathrm{d}t = \mu M$, then the volume increases exponentially, $\mathrm{d}V/\mathrm{d}t = +\mu V$, and the concentration c of a substance (defined as the number of molecules per unit volume) decreases according to $\mathrm{d}c/\mathrm{d}t = -\mu c$.

2. Consider an *E. coli* cell with a generation time of 25 min, a C-period of 40 min and a D-period of 20 min. Using Fig. 5.3 as a guide, show that the newborn cell contains a chromosome with two sets of replication forks, has a replication termination 5 min after birth, and has to initiate at 4 origins 10 min after termination.

6

The eukaryotic cell-cycle engine

The discovery that there is a set of enzymes whose activities correlate with those of cell-cycle events led to the concept of a 'cell-cycle engine'; the enzymes are called cyclin-dependent kinases (CDKs). Current models of the cell cycle focus on the molecular networks regulating the activities of CDKs. Some of these models will be presented in this chapter. After brief sections that provide a basic background on the physiology and biochemistry of eukaryotic cell division, mathematical models of embryonic cell cycles are first discussed because of their simplicity. The regulatory networks of non-embryonic cell cycles are more complex, presumably because they have to be responsive to extracellular and intracellular signals. To highlight the essential network and dynamical elements of the eukaryotic cell-cycle engine, this chapter will illustrate the construction of a cell-cycle model of the budding yeast *Saccharomyces cerevisiae*.

6.1 Physiology of the eukaryotic cell cycle

The eukaryotic cell cycle can be viewed as a mixture of a 'chromosome cycle' and a 'growth cycle'. In a chromosome cycle, DNA is replicated and then segregated between the two daughter cells. In a growth cycle, cellular mass approximately doubles before the cell splits in two. Normally, the chromosome and growth cycles are co-ordinated in eukaryotes, unlike in bacteria where overlapping cell cycles may occur (as illustrated in the preceding chapter).

A schematic diagram of the chromosome cycle is given in Fig. 6.1. DNA is replicated during S phase (S for synthesis of DNA). Chromosomes condense and segregate in M phase (M for mitosis). Gap phases $G1$ and $G2$ separate S and M phases, except in embryonic cell cycles where S and M alternate with no discernible gap phases.

The cell cycle is also sometimes divided into *interphase* (consisting of G1, S, and G2) and *mitosis*. It is during mitosis that the dramatic sequence of events – shown in Fig. 6.2 – is observed. Mitosis is further divided into different subphases, namely, *prophase, metaphase, anaphase, and telophase* (all before the actual process of cell division called *cytokinesis*). In *prophase*, the chromosomes (shown in red in Fig. 6.2) condense, the nuclear membrane breaks down, and the centrosomes migrate to the poles (centrosomes are the star-like entities where spindle fibers, shown in green, emanate). At *metaphase*, each pair of duplicate chromosomes (called 'sister chromatids' that are still bound together) migrates to the equator. The alignment of all the sister chromatids on the equator is a highly regulated process (i.e. separation of sister chromatids does not begin until all of the pairs are aligned on the equator); this alignment is crucial for segregating identical copies of the genome between daughter

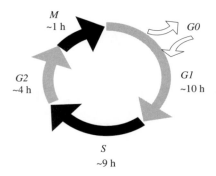

Fig. 6.1 Phases of the eukaryotic cell cycle. Shown are the relative durations of the chromosome cycle phases for an animal cell dividing with a period of 24 h. DNA is replicated in *S* phase. Duplicate chromosomes are segregated in *M* phase (mitosis). *G*1 and *G*2 are 'gap' phases. *G*0 is the quiescent state.

cells. The sister chromatids are separated during *anaphase*. At *telophase* (not shown in Fig. 6.2), new nuclear membranes are formed. The two daughter cells split during *cytokinesis*.

6.2 The biochemistry of the cell-cycle engine

As already mentioned, the cell cycle is driven by a family of enzymes called *cyclin-dependent kinases* (CDKs). These enzymes catalyze the phosphorylation of a variety of proteins involved in processes such as gene expression (e.g. during *S* phase), protein degradation (e.g. during nuclear envelop breakdown in mitosis), replication of centrosomes, segregation of chromosomes, etc. As their names imply, the activation of CDKs absolutely requires binding with proteins called *cyclins*, so-called because of the cyclic variation of their levels during the cell cycle. This variation explains the observed oscillations in the enzymatic activities of the CDKs. Cyclin binding is followed by two phosphorylation events, one due to a kinase called CAK (for *CDK-activating kinase*) that puts an activating phosphate on the CDK and the other due to Wee1 (a tyrosine kinase) that puts an inhibitory phosphate on the CDK. The inhibitory phosphate is removed by a phosphatase called Cdc25. As will be discussed in more detail in Section 6.4, a positive feedback between Cdc25 and CDK exists because the latter activates the former by phosphorylation. Also, note that the active CDK phosphorylates Wee1 cause the inhibition of the latter, and create another positive feedback (of the mutual antagonism type).

There is only one cell cycle CDK in budding yeast (called Cdc28), and also only one in fission yeast (called Cdc2). In contrast, many cyclins have been identified in both yeasts. In mammalian cells, at least nine cell cycle CDKs and at least 20 cyclins have been discovered.

Another group of enzymes considered to be essential drivers of cell-cycle progression are those that target proteins – especially the cyclins – for degradation. For

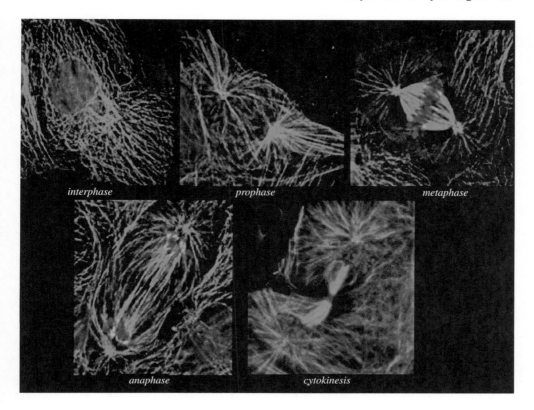

interphase   *prophase*   *metaphase*

anaphase   *cytokinesis*

Fig. 6.2 A dividing cell as seen under a microscope. Microtubules (components of the spindle fibers) are shown in green and chromatin (DNA and associated proteins) is colored red. Photograph courtesy of W.C. Earnshaw, Wellcome Trust Centre for Cell Biology, University of Edinburgh, Scotland, UK. (See Plate 6)

example, as a requirement for anaphase, proteins called cohesins (the glue between sister chromatids) are targeted for degradation by the enzyme complex called APC (for *a*naphase *p*romoting *c*omplex). The APC also targets mitotic cyclins for destruction (inactivating mitotic CDKs as a consequence) in order to exit from mitosis. The APC is a ubiquitin ligase that is an enzyme that catalyzes the transfer of ubiquitin proteins to the target protein (other ubiquitylation enzymes are also involved). The activation of the APC usually requires binding with an activating subunit (such as a protein called Cdc20). Cyclins that are tagged with multiubiquitin chains are then brought into proteasomes for degradation. (A proteasome is a barrel-shaped protein complex where ubiquitinated proteins are degraded into smaller peptides.)

CDK activity is kept low in interphase due to high APC activity. Upon entry into mitosis, APC activity decreases, while CDK activity increases – this observation led Tyson and Novak (2001) to suggest that the essential dynamics of the cell cycle could

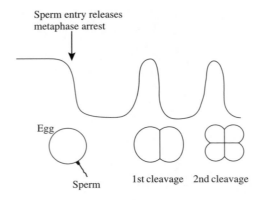

Fig. 6.3 Oscillations in MPF activity during early embryonic cell cycles after fertilization of a frog egg. Twelve synchronous oscillations occur after a sperm enters an egg (Murray and Kirschner, 1989).

be based on the mutual antagonism between mitotic CDKs and APC. This idea will be explained in more detail below.

6.3 Embryonic cell cycles

The maturation-promoting factor (MPF) was discovered in studies of embryonic cell cycles of a frog, *Xenopus laevis* (Murray and Kirschner, 1989). MPF is a protein complex made up of a kinase subunit, CDK1 (also called Cdc2 in the older literature), and a B-type cyclin subunit. Unfertilized frog eggs with high levels of MPF activity are arrested in metaphase. After fertilization by a sperm, MPF activity drops immediately and a series of 12 synchronous MPF oscillations ensues (see Fig. 6.3).

The embryonic cell cycles shown in Fig. 6.3 are alternating S and M phases with no discernible G1 and G2 phases, and no cell growth involved. The cell cycles terminate after twelve divisions. The observation that MPF oscillations also occur in cell-free extracts suggests that an autonomous oscillator is driving embryonic cell cycles.

Early mathematical models of frog embryonic cell cycles – like those of Goldbeter (1991) and of Novak and Tyson (1993) – focused on the following observations: (1) cyclin synthesis is both necessary and sufficient for entry into mitosis, (2) MPF activation is autocatalytic (i.e. MPF enhances its own activation), and (3) cyclin degradation is required for exit from mitosis. The Goldbeter model is discussed in this section, while the Novak and Tyson model will be considered in the next section.

The model proposed by Goldbeter is shown in Fig. 6.4. Cyclin (C) is synthesized at a constant rate, v_i, and degraded through the action of a protease (X). The activation of X is carried out by a cascade of two cyclic enzymatic reactions, namely, the M$^+$-M cycle and the X$^+$-X cycle, where M refers to the active MPF and X to the active protease (the species with a superscript + is the inactive form). Cyclin induces the activation of MPF as indicated by the dashed arrow from cyclin. In return (although

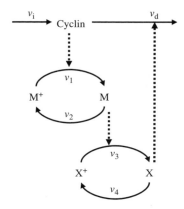

Fig. 6.4 The Goldbeter model of the embryonic mitotic oscillator (Goldbeter, 1991). M^+ and M are inactive and active MPF, respectively. X^+ and X are inactive and active cyclin protease, respectively.

indirectly via X) MPF induces the degradation of cyclin as indicated by the dashed arrow from X. This negative feedback is expected to produce sustained periodic oscillations if there is sufficient delay between the time when cyclin is degraded and the time when X is activated. The delay in the activation of X provided by the cascade of cyclic reactions is required to give the cyclin time to grow before it gets degraded.

To further understand the time delay in the Goldbeter model, one has to consider the dynamical equations (6.1–6.3) and the assumptions made on the values of the parameters. In these equations, C is the cyclin concentration, M is the fraction of the active MPF concentration, and X is the fraction of the active protease concentration.

$$\frac{\mathrm{d}C}{\mathrm{d}t} = v_i - v_d X \frac{C}{K_d + C} - k_d C, \tag{6.1}$$

$$\frac{\mathrm{d}M}{\mathrm{d}t} = V_1(C) \frac{(1 - M)}{K_1 + (1 - M)} - V_2 \frac{M}{K_2 + M}, \tag{6.2}$$

$$\frac{\mathrm{d}X}{\mathrm{d}t} = V_3(M) \frac{(1 - X)}{K_3 + (1 - X)} - V_4 \frac{X}{K_4 + X}, \tag{6.3}$$

where

$$V_1(C) = V_{M1} \frac{C}{K_c + C} \quad \text{and} \quad V_3(M) = V_{M3} M.$$

The factors $(1 - M)$ and $(1 - X)$ are the fractions of the concentrations of inactive MPF and inactive protease, respectively. V_1 to V_4 are the maximum rates of the enzymatic reactions with corresponding numbers in Fig. 6.4. The form of $V_1(C)$ indicates that step 1 is catalyzed by C, and that $V_1(C)$ approaches the maximum value of V_{M1} as C increases; in contrast, no saturation value is imposed for the reaction catalyzed by M as expressed in the form of $V_3(M)$. The last two terms in eqn. 6.4 indicate that the degradation of C is both X-dependent and X-independent.

A key assumption in the Goldbeter model is that both cyclic enzymatic reactions possess the property of *zeroth-order ultrasensitivity*, which requires that the Michaelis constants, K_1 to K_4, are close to zero. This is the case for curves labelled 'a' in Figs. 6.5A and B.

When conditions are met for ultrasensitivity, Fig. 6.5A shows that the cyclin has to pass the threshold value of C* to initiate significant activation of M; similarly, Fig. 6.5B shows that M has to pass the threshold value of M* to induce significant

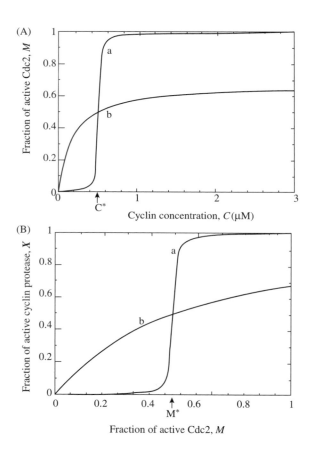

Fig. 6.5 (A) Steady state of the fraction of active MPF (M) as a function of cyclin concentration. (B) Steady state of the fraction of active cyclin protease (X) as a function of M. Curves 'a' in the *top and bottom panels* show zeroth-order ultrasensitivity (when all the Michaelis constants are small, equal to 0.005). Curves 'b' in *both panels* correspond to large values of Michaelis constants (set equal to 10). Figure reproduced with permission from Goldbeter (1991).

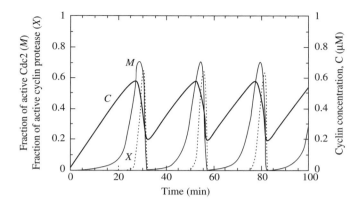

Fig. 6.6 Periodic oscillations exhibited by the Goldbeter model. Parameter values: $v_i = 0.025\,\mu\text{M min}^{-1}$, $v_d = 0.25\,\mu\text{M min}^{-1}$, $K_d = 0.02\,\mu\text{M}$, and $k_d = 0.01\,\text{min}^{-1}$. Initial conditions: $C = 0.01\,\mu\text{M}$, $M = X = 0.01$. Figure reproduced from Goldbeter (1991).

activation of X. The total time taken to reach C* and M* represents sufficient time delay for the negative feedback to generate sustained oscillations, as shown in Fig. 6.6. The character of the oscillations in C and M agree qualitatively with those observed in embryonic cell cycles of frogs.

6.4   Control of MPF activity in embryonic cell cycles

The Goldbeter model does not take into account the experimentally observed auto-catalytic character of MPF activation. Interest in the mechanism of this MPF self-amplification is motivated by observations that it is a target of intracellular signalling pathways that arrest cell-cycle progression – e.g. pathways that emanate from unreplicated DNA or from damaged DNA. Positive-feedback loops that account for MPF self-amplification were considered in a model proposed by Novak and Tyson (1993) for mitotic control in frog embryos. The model's network diagram is shown in Fig. 6.7. Henceforth, this model will be referred to as the *NT93* model.

In Fig. 6.7(a), four phosphorylation states of the CDK subunit (depicted as a rectangular box) of the cyclin/CDK complex are shown (details are given in the figure caption). The complex on the top right corner is the active MPF, while the one on the top left corner is called 'preMPF'. The inactive preMPF is dephosphorylated by the phosphatase Cdc25 to produce active MPF; the tyrosine kinase Wee1 reverses the action of Cdc25. Two positive-feedback loops are created by the actions of Cdc25 and Wee1 as shown in Fig. 6.7(b). The positive feedback between Cdc25 and active MPF is of the mutual-activation type, while the positive feedback between Wee1 and active MPF is of the mutual-antagonism type. A consequence of these positive feedbacks is the potential for bistability. Indeed, *if* total cyclin concentration is considered as a bifurcation parameter (i.e. cyclin synthesis and degradation are temporarily ignored in the NT93 model), the model exhibits bistability as shown in

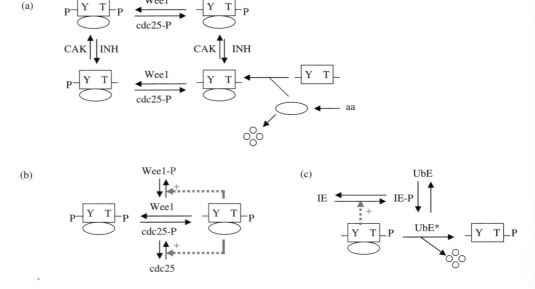

Fig. 6.7 The NT93 model of M-phase control in Xenopus. Figure redrawn from Novak and Tyson (1993). (a) Various phosphorylation states of the cyclin/CDK complex. The *oval figure* is cyclin and the *rectangle* is the CDK. P is a phosphate group. T and Y are threonine and tyrosine amino-acid residues, respectively, in the CDK. aa = amino acids. (b) Details of the phosphorylation and dephosphorylation events catalyzed by Wee1 and cdc25, respectively. (c) Details of the MPF-induced degradation of cyclin. IE = intermediary enzyme; UbE = ubiquitin-conjugating enzyme, CAK = CDK activating enzyme, INH = CDK inhibiting enzyme.

Fig. 6.8. Figure 6.8(a) shows how the steady states of active MPF are approached for different total cyclin concentration levels. The experimentally observed 'cyclin threshold' above which a jump in active MPF occurs can be explained by a saddle-node bifurcation (see Fig. 6.8(b); note that in this figure, the parameter [total cyclin] is the vertical axis). Furthermore, the observed lengthening of time to reach the steady state as the parameter approaches the cyclin threshold (e.g. from [total cyclin] = 0.4 to 0.25 in Fig. 6.8(a)) is consistent with the characteristic slowing down of dynamical trajectories near a saddle-node bifurcation point (recall that, at this bifurcation point, one of the eigenvalues vanishes; the magnitude of eigenvalues indicates how fast the trajectories are moving towards or away from the steady state).

Cyclin synthesis from amino acids and cyclin degradation are shown in Fig. 6.7(a). Details of MPF-induced cyclin degradation are shown in Fig. 6.7(c) (the four small circles depict the products of the degradation of the cyclin subunit of the cyclin/CDK complex) – this portion of the NT93 model is structurally the same as the Goldbeter model because it involves a cascade of two cyclic enzyme reactions (the first cycle

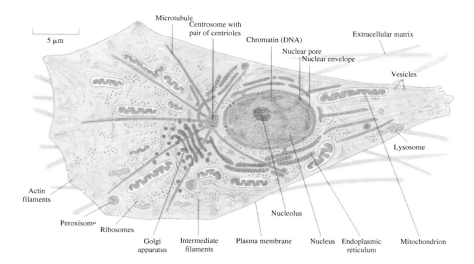

Plate 1

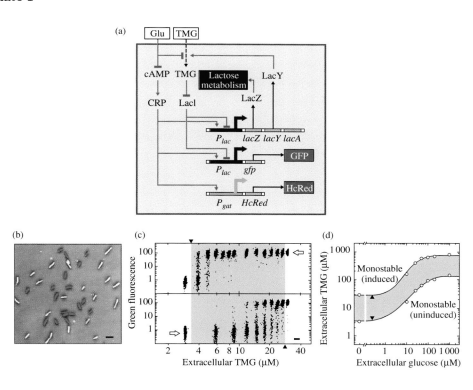

Plate 2

Plate 3

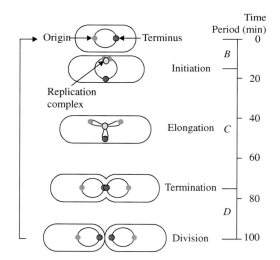

$\tau = 100$ min, $B = 15$ min, $C = 60$ min, $D = 25$ min

Plate 4

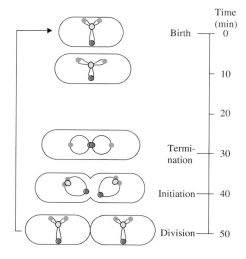

Time
(min)

Birth — 0

— 10

— 20

Termi-
nation — 30

Initiation — 40

Division — 50

$\tau = 50$ min, $C = 40$ min, $D = 20$ min

Plate 5

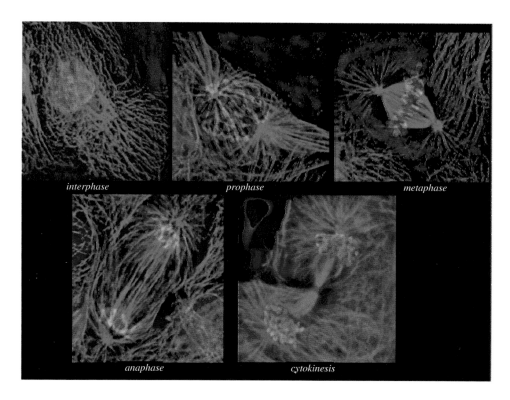

interphase

prophase

metaphase

anaphase

cytokinesis

Plate 6

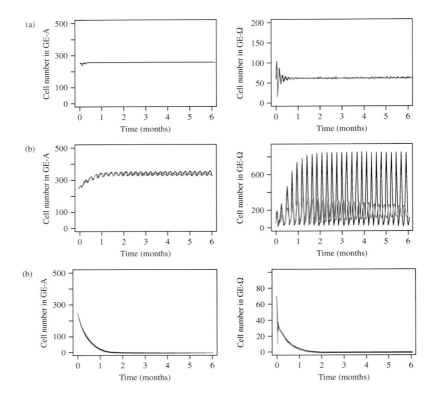

Plate 7

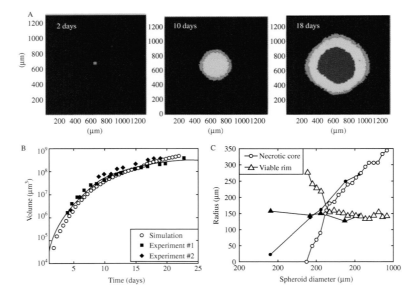

Plate 8

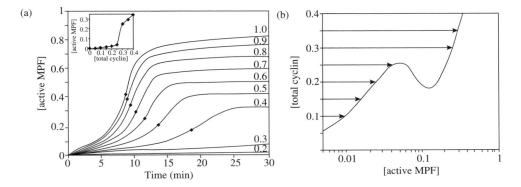

Fig. 6.8 (a) Increase of the activity of MPF with time for various levels of total cyclin (given by the number on top of each curve). The *solid square* on each curve represents 50% of the final (steady-state) [active MPF]. The *inset* shows the final [active MPF] as a function of [total cyclin]. (b) [total cyclin] versus [active MPF] showing the cyclin threshold at about 0.25, above which [active MPF] makes a big jump to a higher level. Figure reproduced with permission from Novak and Tyson (1993).

involves IE and the second involves UbE). As in the Goldbeter model, there is a negative feedback between active MPF and UbE*, the latter causing the degradation of the cyclin subunit of MPF. The oscillations generated by this negative feedback are reminiscent of those observed in intact frog embryos.

6.5 Essential elements of the basic eukaryotic cell-cycle engine

A stepwise construction of a cell-cycle model network of the budding yeast, *S. cerevisiae*, will be illustrated in this section. The discussion closely follows that of the paper of Tyson and Novak (2001). The recent publication of Csikasz-Nagy *et al.* (2006) attempts a comprehensive integration of the modelling performed by the Novak–Tyson group over many years of the cell cycles of various organisms – including budding yeast, fission yeast, Xenopus embryo, and mammalian cells.

In budding yeast, the only cell cycle CDK is cdc28; but there are several cyclins that can bind cdc28 to form complexes that function in different cell-cycle phases. The model below focuses on only one cyclin, namely, a B-type cyclin (symbolized by cycB in the equations; it is a cyclin that belongs to the so-called *Clb* family). The model construction starts with the premise that the mutual antagonism between the CDK and APC (anaphase-promoting complex) is the ultimate origin of two 'self-maintaining' states: the G1 state (characterized by high APC and low CDK activities), and the S-G2-M state (characterized by low AP and high CDK activities). Two irreversible transitions, called *start* and *finish*, delineate one state from the other. These basic features are depicted in Fig. 6.9. Sufficient CDK activity is required to trigger the *start* transition, and sufficient APC activity is needed for the *finish* transition. Coupled

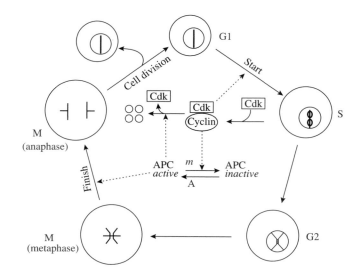

Fig. 6.9 The various phases of the eukaryotic cell cycle (G1, S, G2, M) and the position of the irreversible transitions *Start*, and *Finish*. Active Cdk/Cyclin triggers *Start*, while active APC (analphase-promoting complex) triggers *Finish*. The antagonism between Cdk/Cyclin and APC is shown. The Cdk-induced inhibition of the APC increases with cell growth (increasing mass, *m*). The activation of APC is induced by an unidentified factor *A*. Figure reproduced with permission from Tyson and Novak (2001).

with the growth of the cell (i.e. increasing cell mass, m), an increase in CDK activity leads to inhibition of APC. On the other hand, increasing APC activity increases the rate of cyclin degradation and therefore the inhibition of the CDK.

The dynamical variable used to represent CDK activity is the concentration of the cyclin, $[cycB]$. It can be assumed that CDK activity is proportional to the cyclin concentration because cellular levels of the free CDK subunit are observed to be non-rate limiting and approximately constant. The variable representing APC activity is $[Cdh1]$, which refers to the concentration of the auxiliary protein Cdh1 that activates APC. For the simple network shown in Fig. 6.9, one can write the following kinetic equations:

$$\frac{\mathrm{d}[cycB]}{\mathrm{d}t} = k_1 - k_2'[cycB] - k_2''[Cdh1][cycB], \tag{6.4}$$

$$\frac{\mathrm{d}[Cdh1]}{\mathrm{d}t} = \frac{k_3'(1-[Cdh1])}{J_3 + (1-[Cdh1])} + \frac{k_3''A(1-[Cdh1])}{J_3 + (1-[Cdh1])} - \frac{k_4 m[cycB][Cdh1]}{J_4 + [Cdh1]}. \tag{6.5}$$

The cyclin is synthesized at a constant rate k_1, and degraded without or with the influence of Cdh1 (second and third terms on the right-hand side of eqn 6.4. In eqn 6.5, the fraction of active Cdh1 is given by *[Cdh1]* and the fraction of the inactive form is *(1−[Cdh1])*. The conversion from the inactive to the active form of Cdh1 occurs with or without the catalysis of an enzyme *A* (the identity and significance of which

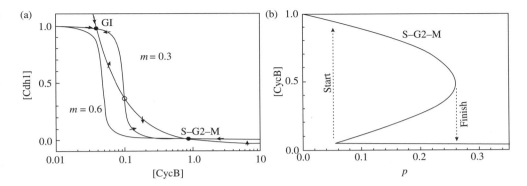

Fig. 6.10 (a) The nullclines of eqns 6.4 and 6.5. The Cdh1 nullcline is drawn for two values of the mass m. Parameter values are $A = 0$; rate constants (min^{-1}): $k_1 = 0.04$, $k_2' = 0.04$, $k_2'' = 1$, $k_3' = 1$, $k_3'' = 10$, $k_4 = 35$, $k_4' = 2$; Michaelis constants (dimensionless): $J_3 = J_4 = 0.04$. (b) Steady state of cycB as a function of the parameter p where $p = (k_3' + k_3''A)/k_4 m$. Figures reproduced with permission from Tyson and Novak (2001).

will be discussed later) according to the second and first terms, respectively, of the right-hand side of eqn 6.5. The last term in eqn 6.5 assumes that the inhibition of Cdh1 is catalyzed by cycB and that this rate increases with cell mass m. Tyson and Novak assume that the inactivation of Cdh1 by the CDK occurs in the nucleus where cycB accumulates; thus, $[cycB]$ is multiplied by m to account for the increase in the effective concentration of cycB as the cell grows.

The nullclines from eqns 6.4 and 6.5 are plotted in Fig. 6.10(a); these demonstrate how an increase in mass ($m = 0.3$ to 0.6) switches the system from the G1 stable steady state to the S-G2-M stable steady state. Note that the system can have 1 or 3 steady states depending on the mass. Figure 6.10(b) shows a plot of the $[cycB]$ steady states as a function of the lumped parameter p whose expression arises naturally from eqn 6.5. The *start* and *finish* transitions are associated with saddle-node bifurcation points. Starting from the G1 branch of the steady-state curve, as the cell mass increases (equivalent to decreasing p), the system will eventually reach the left knee of the curve where it makes the *start* transition to the upper branch of the steady-state curve if mass continues to grow. From the S–G2–M state, the system can move towards the right knee of the curve if p increases (which is implemented by an abrupt increase in the parameter A – a situation discussed below). After the *finish* transition, the cell splits in two and the mass per cell is half of what it was at the transition.

At this point, the model represented by eqns 6.4 and 6.5 does not generate CDK oscillations. Other molecular steps have to be considered to enable the system to switch between the lower and upper branches of the steady-state curve (these are called hysteretic oscillations). In fact, a pathway for CDK-induced activation of APC has been found; it is a pathway that involves the CDK-induced activation of an APC-activating factor called cdc20 activates Cdh1 (see Fig. 6.11). The activity of cdc20, symbolized

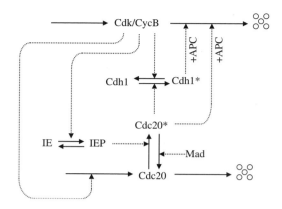

Fig. 6.11 The simple model of Fig. 6.9 is augmented to include another APC-activating factor, Cdc20, and an intermediary enzyme (IE). Figure redrawn from Tyson and Novak (2001).

by $[Cdc20_A]$, therefore replaces the parameter A in eqn 6.5, and is now considered an additional dynamical variable. Cdc20 exists in both inactive and active forms, and the model assumes that the inactive form is continuously synthesized in a CDK-dependent manner. As in the Goldbeter model and the NT93 model for embryonic cell cycles, a time delay between APC-induced degradation of cycB and activation of APC (i.e. activation of Cdh1 and Cdc20) is provided by the cascade of cyclic enzyme reactions, including an assumed intermediary enzyme IE that activates Cdc20. The network shown in Fig. 6.11, excluding Cdh1 and ignoring mass m as a parameter, can generate embryonic cell-cycle oscillations similar to the Goldbeter model discussed in Section 6.3.

The mass m, a parameter in eqn 6.5, is now also considered a dynamical variable. The additional dynamical equations are

$$\frac{d[Cdc20_T]}{dt} = k_5' + k_5'' \frac{(m[cycB])^n}{J_5^n + (m[cycB])^n} - k_6[Cdc20_T], \tag{6.6}$$

$$\frac{d[Cdc20_A]}{dt} = \frac{k_7[IEP]([Cdc20_T] - [Cdc20_A])}{J_7 + ([Cdc20_T] - [Cdc20_A])} - \frac{k_8[Mad][Cdc20_A]}{J_8 + [Cdc20_A]} \tag{6.7}$$
$$- k_6[Cdc20_A],$$

$$\frac{d[IEP]}{dt} = k_9 m[cycB](1 - [IEP]) - k_{10}[IEP], \tag{6.8}$$

$$\frac{dm}{dt} = \mu m \left(1 - \frac{m}{m_*}\right), \tag{6.9}$$

where $[Cdc20_T]$ is the total concentration of Cdc20, $[Cdc20_A]$ is the concentration of the active form of Cdc20, and $[IEP]$ is the fraction of the phosphorylated (active) form of IE. In eqn 6.9, m_* is the maximum mass that the cell can grow to if it does not divide. The protein Mad inhibits Cdc20 as shown in Fig. 6.11; it is a fixed parameter and not considered a dynamical variable. Mad is a family of *spindle-checkpoint* proteins that inhibits Cdc20 to prevent the APC from initiating anaphase if not all the sister chromatids are aligned on the metaphase plane (to be discussed more in the next chapter).

The model, now consisting of eqns 6.4 (with A replaced by Cdc20) to 6.9, generates the periodic oscillations shown in Fig. 6.12. These oscillations are not autonomous in the sense that the mass is reduced to half 'whenever the cell divides' which is a decision

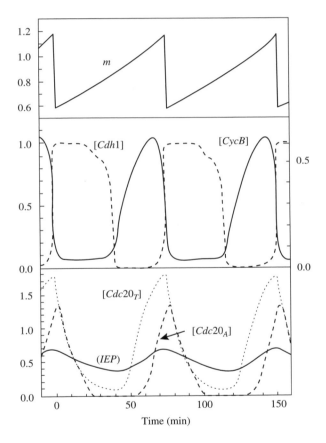

Fig. 6.12 Periodic oscillations generated by eqns 6.4 to 6.9. When $[CycB_T]$ crosses the threshold of 0.1 from above, cell division is assumed to occur and cell mass is halved. Parameter values as in Fig. 6.10(a) except, now, the factor A is replaced by Cdc20 in Fig. 6.11 with the following associated parameters: (rate constants, $\min^{-1}$) $k_5' = 0.005$, $k_5'' = 0.2$, $k_6 = 0.1$, $k_7 = 1$, $k_8 = 0.5$, $k_9 = 0.1$, $k_{10} = 0.02$, (Michaelis constants, dimensionless) $J_5 = 0.3$, $J_7 = 0.001$, $J_8 = 0.001$; other dimensionless parameters: $n = 4$, $[Mad] = 1$. Figures reproduced with permission from Tyson and Novak (2001).

that is imposed by the modeller (in the case of Fig. 6.12, this decision to divide is made whenever $[cycB]$ drops below 0.1).

CDK activity is also regulated by a family of proteins called CKIs (for CdK-Inhibitors) that bind cyclin/CDK complexes to form trimers that lack kinase activity. An example of a CKI in budding yeast is Sic1. High levels of Sic1 prevent entry into the cell cycle by keeping cycB/CDK activity low. However, cycB/CDK can phosphorylate Sic1, which leads to a pathway towards Sic1 degradation. To help cycB/CDK, the initial inhibition of Sic1 is carried out by a starter kinase (SK). In budding yeast, the SK is a Cln/CDK complex where Cln is another type of cyclin that is different from cycB. (In budding yeast there are several Cln cyclins and Clb cyclins; cycB in the model is a member of the family of Clb cyclins.) This picture completes the basic cell-cycle engine of the budding yeast. Novak and Tyson (2001) go further by proposing that the elements of this cell-cycle network characterize all eukaryotic cell cycles, including mammals. The full model is shown in Fig. 6.13. The full set of equations is composed of eqns 6.6 to 6.9 above, eqns 6.10 to 6.11 below, and the modificatiions of eqns 6.4 to 6.13 and 6.6 to 6.14 below. The modification in eqn 6.14 includes the assumption

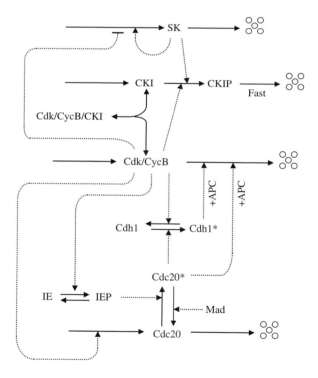

Fig. 6.13 The basic eukaryotic cell-cycle engine according to Tyson and Novak. SK = starter kinase, CKI = cdk-inhibitor. Figure redrawn from Tyson and Novak (2001).

that SK phosphorylates Cdh1 (not shown in Fig. 6.13).

$$\frac{d[CKI_T]}{dt} = k_{11} - (k_{12}' + k_{12}''[SK] + k_{12}'''m[cycB])[CKI_T], \tag{6.10}$$

$$\frac{d[SK]}{dt} = k_{13}[TF] - k_{14}[SK], \tag{6.11}$$

where

$$[TF] = G(k_{15}'m + k_{15}''[SK], k_{16}' + k_{16}''[cycB], J_{15}, J_{16}), \tag{6.12}$$

$$\frac{d[cycB_T]}{dt} = k_1 - (k_2' + k_2''[Cdh1] + k_2'''[Cdc20_A])[cycB_T], \tag{6.13}$$

where $[cycB_T] = [cycB] + [Trimer]$ with *Trimer* referring to the cycB/CDK/CKI complex;

$$\begin{aligned}\frac{d[Cdh1]}{dt} &= \frac{k_3'(1 - [Cdh1])}{J_3 + (1 - [Cdh1])} + \frac{k_3''[Cdc20_A](1 - [Cdh1])}{J_3 + (1 - [Cdh1])} \\ &- \frac{(k_4'[SK] + k_4 m[cycB])[Cdh1]}{J_4 + [Cdh1]}.\end{aligned} \tag{6.14}$$

$[CKI_T]$ is the total concentration of the CKI and $[cycB_T]$ is the total concentration of cycB.

The CKI is synthesized at a constant rate of k_{11}, and it is degraded at a basal rate (associated with rate constant k_{12}') and at rates that depend on $[SK]$ and on $m[cycB]$. The synthesis of SK depends on the concentration of transcription factors, $[TF]$, which is given by the Goldbeter–Koshland function G in eqn 6.12. The autocatalytic nature of SK activation and the inhibition of SK by cycB/CDK are encoded in the arguments of the function G (see top of Fig. 6.13).

Novak and Tyson (2001) validate their model using observations from gene knock-out experiments and deletion mutants. Computer simulation using the model is shown in Fig. 6.14 (note that the full set of equations is associated with the 'wild-type' phenotype). Simulation of *Cln* mutants (e.g. the triple mutant *cln1Δ cln2Δ cln3Δ*) by setting $k_{13} = 0$ is shown in Fig. 6.14(b), which agrees with the observation that these mutant cells arrest in G1 with high CKI and Cdh1 activities. Simulation of mutant cells lacking SK and CKI (e.g. the quadruple mutant *cln1Δ cln2Δ clnΔ3 sic1Δ*), by setting $k_{11} = k_{13} = 0$, is shown in Fig. 6.14(c), which is consistent with the observation that these mutants are viable (it is thought that the only essential role of the *Cln*-cyclins is to remove the CKI).

6.6 Summary

The concept of an 'engine' in the form of an autonomous CDK oscillator that orches-trates the sequence of complex cell-cycle events is an attractive one. Studies of early embryonic cell cycles in frog (Xenopus) provide evidence of the existence of such an

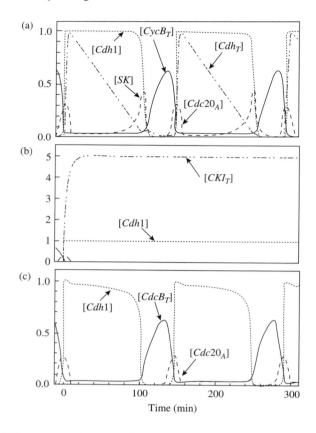

Fig. 6.14 (a) Wild-type simulation of the full budding yeast cell cycle corresponding to eqns 6.6 to 6.11, 6.13 and 6.14. Parameter values: all those used in Fig. 6.12 and (rate constants, min^{-1}) $k_{11} = 1$, $k'_{12} = 0.2$, $k''_{12} = 50$, $k'''_{12} = 100$, $k_{13} = 1$, $k_{14} = 1$, $k'_{15} = 1.5$, $k''_{15} = 0.05$, $k'_{16} = 1$, $k''_{16} = 3$, $\mu = 0.005$, and (dimensionless parameters) $K_{eq} = 10^3$, $J_{15} = J_{16} = 0.01$, $m_* = 10$. (b) Mutants with no SK are simulated by setting $k_{13} = 0$. (c) Mutant cells without SK and CKI are simulated by setting $k_{11} = k_{13} = 0$. Figures reproduced with permission from Tyson and Novak (2001).

oscillator. The Goldbeter model for embryonic cell cycles focuses on the role of a negative feedback between cyclin and a cyclin protease in generating CDK oscillations. In addition to this negative feedback, the Novak–Tyson (NT93) model incorporates mechanistic details of the CDK self-amplification (positive feedback) and predicts bistable behavior.

A detailed construction of a non-embryonic cell-cycle model, using the budding yeast *S. cerevisiae* as an example, was illustrated using the work of Tyson and Novak (2001). In this model, the coupling between cell growth and CDK activation is accounted for. It should be noted, however, that the nature of this coupling is an open problem at this time. The basic premise of the Novak–Tyson yeast model discussed at the end of this chapter is that the mutual antagonism (i.e. a positive feedback) between the CDK and the anaphase-promoting complex (APC) generates two stable steady states corresponding to the so-called 'self-maintaining states', namely, G1 and S-G2-M. The positive feedback in this yeast model generates the potential for bistability under certain conditions; and the coupling between this positive-feedback with a negative-feedback loop in the network is ultimately the origin of hysteretic oscillations exhibited by the model.

References

Csikasz-Nagy, A., Battogtokh, D., Chen, K. C., Novak, B., and Tyson, J. J., (2006) 'Analysis of a generic model of eukaryotic cell-cycle regulation,' *Biophysical Journal* **90**, 4361–4379.

Goldbeter A. (1991) 'A minimal cascade for the mitotic oscillator involving cyclin and cdc2 kinase,' *Proceeding of National Academy of Sciences of the USA* **88**, 9107–9111.

Murray, A. W. and Kirschner, M. W. (1989) 'Dominoes and clocks: The union of two views of the cell cycle,' *Science* **246**, 614–621.

Novak, B. and Tyson, J. J. (1993) 'Numerical analysis of a comprehensive model of M-phase control in Xenopus oocyte extracts and intact embryos,' *Journal of Cell Science* **106**, 1153–1168.

Tyson, J. J. and Novak, B. (2001) 'Regulation of the eukaryotic cell cycle: molecular antagonism, hysteresis, and irreversible transitions,' *Journal of Theoretical Biology* **210**, 249–263.

Exercises

1. The Goldbeter model (Fig. 6.4 and eqns 6.1–6.3) does not consider any positive feedback between the dynamical variables; for this reason, one could expect that the model does not exhibit bistability. Verify that the Goldbeter model does not possess multiple steady states for any given set of parameters (particularly for parameter values near those used in Fig. 6.6.

2. Show that the *start* and *finish* transitions in Fig. 6.10(b) correspond to saddle-node bifurcation points, and find the corresponding values of the parameter p at these points. Analyze the linear stability of the different branches of the steady-state curve shown in Fig. 6.10(b).

3. The time for a given initial cell mass to double is controlled by the parameter μ (see eqn 6.9). There is a sensitive relation between μ and the $[cycB]$ threshold that signals when the cell mass is halved (see caption of Fig. 6.12). By performing computer simulations similar to those of Fig. 6.12, find another pair of values for μ and $[cycB]$ threshold that gives periodic oscillations.

7

Cell-cycle control

Signalling pathways are said to control the cell-cycle engine discussed in the preceding chapter by regulating transitions between phases in the cell cycle. These are transitions where putative decisions are made such as initiating DNA replication (entry into S phase) or segregating duplicated chromosomes (entry into M phase). The mechanisms for regulating these transitions involve the so-called *cell-cycle checkpoints*. Modelling of a G1 checkpoint called the *restriction point* in mammalian cells is illustrated in detail in this chapter; it is a checkpoint where a cell's commitment to DNA replication is made. Also discussed is a G2 checkpoint that prevents cell-cycle progression into mitosis when, for example, DNA damage is not repaired. The segregation of sister chromatids at anaphase is guarded by the so-called *metaphase checkpoint* (also called the mitotic *spindle checkpoint*); it checks that all the chromosomes are aligned on the metaphase plate prior to anaphase.

7.1 Cell-cycle checkpoints

Cell-cycle checkpoints are points in the cell cycle where decisions are made whether cell-cycle progression continues or halts. The 'point' in checkpoint is a source of confusion in the literature. Intuitively, a cell-cycle checkpoint involves a surveillance mechanism that somehow checks whether the requirements for progression to the next cell-cycle phase are satisfied and, if not, a mechanism is triggered to arrest the process. These checkpoints are classified as either 'intrinsic' or 'extrinsic' (Elledge, 1996). To illustrate, Fig. 7.1(a) shows two sequences of cell-cycle events arising from 'a', namely, a→b→c and a→d→e. The pathway with a hammerhead (indicating inhibition) and the lifting of this inhibition by 'c' is an example of an intrinsic checkpoint mechanism; it ensures that event 'e' does not occur before 'c'. The pathway labelled 'extrinsic' in Fig. 7.1(a) representing, for example, a signal from DNA damage is an extrinsic checkpoint pathway that arrests the cell cycle at the indicated point.

In Fig. 7.1(b), examples of the major cell-cycle checkpoints are shown. The DNA replication and the spindle assembly checkpoints are examples of intrinsic check-points – the first ensures that the G2→M transition (prophase) is blocked until all the DNA are replicated, while the second prevents the metaphase→anaphase transition until all of the sister chromatids are aligned properly at the metaphase plate. Extrinsic DNA damage checkpoints are shown in the figure as operating at the G1→S and G2→M transitions as well as during S phase. Cell-cycle progression is arrested at these points presumably to give DNA damage-repair programs time to act. The G1

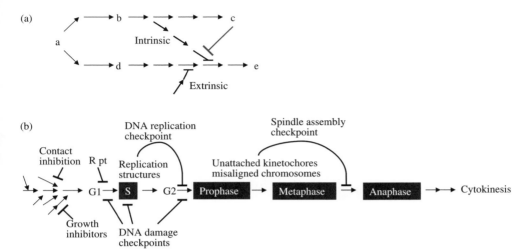

Fig. 7.1 (a) Illustration of intrinsic and extrinsic checkpoints. (b) Cell-cycle checkpoints: Restriction point (R pt), DNA damage checkpoints, DNA replication checkpoint, and spindle assembly checkpoint. Figures are redrawn from Elledge (1996).

and G2 checkpoints, in addition to their role of checking the integrity of the DNA, also ensure that the cell has grown to the appropriate size and has enough nutrients to carry out the next cell-cycle phase.

7.2 The restriction point

There are at least nine CDKs and at least twenty cyclins involved in the mammalian cell cycle. A typical temporal pattern in the activities of some of the major cyclin/CDK complexes is shown in Fig. 7.2.

As shown in Fig. 7.2(a), the major cyclins belong to types A, B, D, and E. The D-type cyclins are often referred to as 'growth-factor sensors' because their expression is upregulated in response to growth factors. These cyclins specifically bind CDK4 and CDK6. The activity of cyclin E/CDK2 increases just before S phase and declines in S phase. CyclinA/CDK2 activity increases progressively during S-G2. The rapid increase in cyclin B/CDK1 activity occurs during the G2-M transition and a sharp drop is required for exit from mitosis.

The G1 checkpoint is referred to as the restriction point (or *R point*) in mammalian cells, and is said to be a 'commitment point' for DNA replication. Quiescent cells exposed to growth factors for at least a minimum amount of time are committed to enter S phase – in other words, withdrawal of growth factors after this minimum time does not stop DNA replication, but withdrawal before this time does. As an operational definition, the R point is the point in time after which withdrawal of growth factors does not stop entry into S phase. (The R point is analogous to the *Start* transition in yeast discussed in the preceding chapter.)

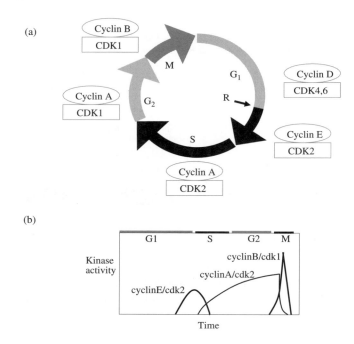

Fig. 7.2 (a) Major cyclin/CDK complexes and where their peak activities occur during the mammalian cell cycle. R indicates the position of the 'restriction point' in mid- to late-G1. (b) Temporal variation of the cyclin/CDK activities during the mammalian cell cycle. Figure 7.2(b) is redrawn from Pines (1999).

The experimental determination of the position of the R point is shown schematically in Fig. 7.3. The accumulation of cyclin E protein and the concomitant activation of CDK2 are the molecular markers for S-phase entry. It has been demonstrated experimentally that the position of the R point in mammals is located between 3 and 4 h after the previous mitosis (Pardee, 1974). Note that cyclin E levels are very low (if not absent) immediately after the R point. Also, interestingly, most of the cyclin E protein is degraded within 1–2 h after S-phase entry.

The control of the R point is of medical interest because of the fact that almost all known human cancers involve malfunctions of this checkpoint. In the next section, the complex regulatory network of the G1–S transition in mammalian cells will be presented and an approach to extracting a model of the R point will be discussed.

7.3   Modelling the restriction point

7.3.1   The G1–S regulatory network

A model network of the G1–S transition in mammalian cells is shown in Fig. 7.4. S-phase entry is induced by growth factor (GF)-mediated signalling pathways that lead to the activation of the pre-replication complex (pre-RC) – a group of proteins

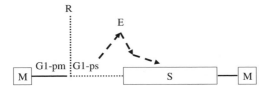

Fig. 7.3 This figure shows that the R point splits the G1 phase into the G1-pm (post-mitosis) period of relatively fixed length (3–4 h), and the G1-ps (pre-S phase) of variable length (1–10 h) associated with the accumulation of cyclin E (symbolized as E in the figure). Upon entry into S phase, most of the cyclin E protein is degraded within 1–2 h. Figure is adapted from Ekholm *et al.* (2001).

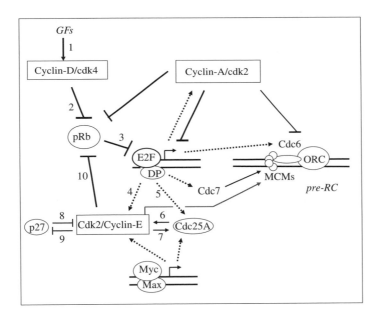

Fig. 7.4 Pathways from growth-factor signalling to the pre-replication complex (pre-RC) in the G1–S transition of the mammalian cell cycle. See text for discussion. The pre-RC forms at replication origins on the DNA (shown in the figure as a duplex of horizontal lines). GFs = growth factors, pRb = retinoblastoma protein, ORC = origin recognition complex. *Arrows* mean 'activate', and *hammerheads* mean 'inhibit'. *Dashed arrows* symbolize gene expression. *Solid lines* are post-translational modifications.

that possesses enzymatic activity for unzipping the DNA duplex at the origins of replication, the key first step in DNA replication.

In quiescent cells, S-phase is blocked by the retinoblastoma protein (pRb) by binding and inhibiting the E2F/DP transcription factors. The E2Fs form heterodimers with DP proteins, and together they induce expression of many genes (some of which are shown by the dashed arrows in Fig. 7.4) that promote S-phase entry; these genes include cyclin E, cyclin A, the phosphatase Cdc25A, and proteins involved in the formation or activation of the pre-RC (e.g. Cdc6 and Cdc7). In addition to pRb, various CDK inhibitors (CKIs) reinforce the quiescent state. These proteins belong to either the Cip/Kip or the INK4 families; p27^{Kip1} (shown in Fig. 7.4) is an example of the first family, and p16^{INK4A} is an example of the second. Another significant transcription factor that induces S-phase entry is c-Myc; as shown in Fig. 7.4, c-Myc shares common transcriptional targets with the E2Fs.

Growth factors stimulate signalling pathways that upregulate the synthesis of D-type cyclins. These cyclins bind and activate CDK4 or CDK6. These so-called G1 CDKs phosphorylate and inactivate pRb, thereby allowing the expression of S-phase genes including cyclins E and A; these cyclins activate CDK2 that further phosphorylates and inactivates pRb. Positive-feedback loops are thus created that accelerate the inactivation of pRb, freeing up E2F/DP to induce expression of S-phase genes. The inhibitory action of cyclin A/CDK2 against the E2F/DP transcription factors (by phosphorylation of the latter) and on the pre-RC have been suggested as possible mechanisms for ensuring that the chromosomes are replicated only once per cell cycle.

Since cyclin E/CDK2 activation is considered to be the marker for S-phase entry (see Fig. 7.3), its regulation is the focus of modelling the R point. Besides its pRb-mediated transcriptional repression, cyclin E/CDK2 is also inhibited by p27^{Kip1} (in fact, a mutual antagonism exists – see the interactions numbered 8 and 9 in Fig. 7.4). This CKI forms inactive trimers with cyclin E/CDK2, and CDK2 phosphorylates p27^{Kip1} leading to the latter's degradation.

Another protein that regulates cyclin E/CDK2 activity is the phosphatase Cdc25A. Just like the autocatalytic activation of MPF in the G2-M transition discussed in the previous chapter, a positive-feedback loop between Cdc25A and CDK2 exists (see interactions numbered 6 and 7 in Fig. 7.4). Cdc25A activates CDK2 by removing an inhibitory phosphate and, in return, CDK2 activates Cdc25A by phosphorylation.

7.3.2 A switching module

To begin to understand the kinetic origin of the R point, one investigates the G1–S network for possible sources of switching behavior. The subnetwork shown in Fig. 7.5 represents a good candidate mechanism that can explain the origin of the switching behavior of cyclin E/CDK2 at the G1–S transition. Mechanistic details of the qualitative interactions depicted in Fig. 7.5(a) are given in Fig. 7.5(b). Note that there are two positively coupled phosphorylation-dephosphorylation (PD) cycles involving cyclin E/CDK2 and Cdc25A. Many coupled PD cycles are present in the regulation of cell-cycle events (PD networks are also referred to as kinase–phosphatase networks).

Regardless of the kinetics of the positively coupled PD cycles involving CDK2 and Cdc25A, one can show that a switching behavior occurs for some parameter values (Aguda, 1999a). This switching behavior is due to a transcritical bifurcation, as explained in Fig. 7.6.

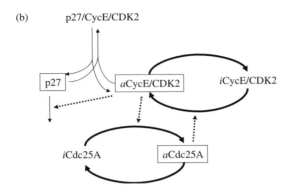

Fig. 7.5 (a) A subnetwork involving CDK2 and its regulators, the CKI called p27 and the phosphatase Cdc25A. (b) The same subnetwork as in (a) with known details of the mechanism. Active cyclin E/CDK2 (*a*CycE/CDK2) is inhibited by p27 by trimer formation, and *a*CycE/CDK2 phosphorylates p27 inducing the latter's degradation. Inactive cyclin E/CDK2 (*i*CycE/CDK2) is activated by active phosphatase *a*Cdc25A that is activated in return by *a*CycE/CDK2 (by phosphorylation).

Using the mass-action kinetic equations given in the caption of Fig. 7.6, one can show that the transcritical bifurcation point is given by the following relationship:

$$(E_1 E_2)^* = \left(\frac{k_{1r}}{k_{1f}}\right)\left(\frac{k_{2r}}{k_{2f}}\right) \tag{7.1}$$

As illustrated in Fig. 7.6(c), the steady states of Y_1 and Y_2 both become positive only if $E_1 E_2 > (E_1 E_2)^*$. Applying this result to the coupled cycles involving cyclin E/CDK2 and Cdc25A, one can claim that the total protein levels (active plus inactive) of cyclin E/CDK2 and Cdc25A must increase in order for the activities of CDK2 and Cdc25A to turn on (become positive). Importantly, coupling the CDK2-Cdc25A switch with CDK2-p27 mutual antagonism (see Fig. 7.6) generates a sharp switching dynamics for the activation of cyclin E/CDK2. The module shown in Fig. 7.6 is the ultimate origin of the switching behavior of cyclin E/CDK2 in the R point model of Aguda and Tang (1999).

7.4 The G2 DNA damage checkpoint

A switching module with a structure similar to Fig. 7.6(a) can be found in the G2-M regulatory network of the mammalian cell cycle; this is shown in the bottom layer of Fig. 7.7 and consists of Wee1, MPF, and Cdc25C. The so-called 'G2 DNA damage

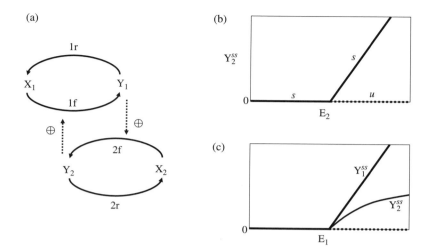

Fig. 7.6 (a) Two positively coupled cyclic reactions. The reaction rates are: $v_{1f} = k_{1f}[X_1][Y_2]$, $v_{1r} = k_{1r}[Y_1]$, $v_{2f} = k_{2f}[X_2][Y_1]$, $v_{2r} = k_{2r}[Y_2]$. A set of two independent kinetic equations are: $d[Y_1]/dt = v_{1f} - v_{1r}$, $d[Y_2]/dt = v_{2f} - v_{2r}$ with $[X_1] = E_1 - [Y_1]$ and $[X_2] = [E_2] - [Y_2]$, where E_1 and E_2 are constants. $[Y_1]_{ss}$ and $[Y_2]_{ss}$ are steady-state concentrations of Y_1 and Y_2, respectively. For graphs (b) and (c), *solid curves* are locally stable steady states (s) and *dotted lines* are unstable steady states (u). Superscript *ss* means 'steady state'.

checkpoint' (G2DDC) system shown in Fig. 7.7 involves signalling pathways emanating from sensors of DNA damage and ending at the Wee1-MPF-Cdc25C module. Only the essential features of the operation of the G2DDC will be discussed here (mechanistic details can be found in Aguda (1999b)).

Recall that MPF is identical to cyclin B/CDK1. Some mechanistic details of the mutual antagonism between Wee1 and MPF, and the mutual activation between Cdc25C and MPF, are shown in Fig. 7.8. The positive-feedback loop between MPF and Cdc25C is similar to the interaction between cyclin E/CDK2 and Cdc25A shown in Fig. 7.5(b). However, the mutual antagonism between Wee1 and MPF is different from the mutual antagonism between p27Kip1 and CDK2, as shown in the upper half of Fig. 7.8.

The nature of the instability of the MPF-Cdc25C module in the G2DDC is similar to that of the CDK2-Cdc25A module in the R point; both checkpoints exhibit transcritical bifurcation when the protein levels cross the values according to eqn 7.1. The Wee1-MPF-Cdc25C module in the G2DDC system is expected to generate a sharp switching behavior similar to the one generated by the p27Kip1-CDK2-Cdc25A module in the R point network (see Fig. 7.5(a)); indeed, this sharp switch is demonstrated by the computer simulations shown in Fig. 7.9. Note that the activities of Cdc25 and MPF become positive at the same time – a hallmark of a transcritical bifurcation in positively coupled cyclic reactions (see Fig. 7.6(c)).

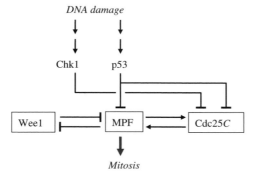

Fig. 7.7 DNA-damage-signalling pathways via the kinase Chk1 and the tumor-suppressor protein p53. *Arrows* mean activate or upregulate, and *hammerheads* mean inhibit or down-regulate. Wee1 is a tyrosine kinase and Cdc25C is a phosphatase. Figure adapted from Aguda (1999b).

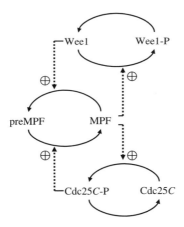

Fig. 7.8 Coupled phosphorylation–dephosphorylation cycles involving MPF (cyclin B/ CDK1), Wee1, and Cdc25C. The network corresponds to the lower part of Fig. 7.7 showing the mutual antagonism between Wee1 and MPF, and the mutual activation between Cdc25C and MPF. Figure adapted from Aguda (1999a).

   Recently, the DNA-damage-signalling pathways that arrest the cell cycle in G1 have been elucidated. A summary of these pathways is given in Fig. 7.10. From a comparison between Figs. 7.7 and 7.10, one may suggest the conclusion that DNA damage-checkpoint-signalling pathways target the cell-cycle-specific CDK-Cdc25 couple that possesses an intrinsic instability (transcritical bifurcation) that enables checkpoint signalling to switch cell-cycle progression on or off.

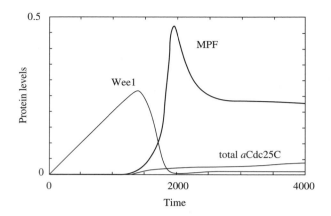

Fig. 7.9 Computer simulations using the detailed G2 DNA-damage-checkpoint model of Aguda (1999b). Figure reproduced with permission from Aguda (1999b). Copyright 1999 National Academy of Sciences, USA.

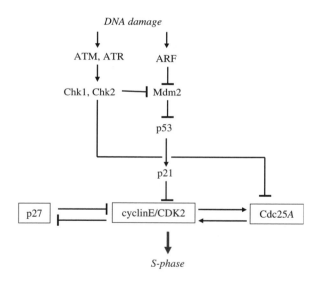

Fig. 7.10 G1 DNA-damage-checkpoint pathways (Bartek and Lukas, 2001). Compare with the G2 DNA-damage pathways shown in Fig. 7.7.

7.5 The mitotic spindle checkpoint

After DNA replication in S phase, the chromosomes condense and the sister chromatids are drawn to the midplane between the centrosomes in preparation for anaphase (see Fig. 6.2). Spindle fibers (microtubules) emanating from the polar centrosomes elongate

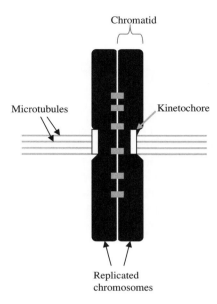

Fig. 7.11 Sister chromatids, or replicated chromosomes still glued together by cohesins (*small grey rectangles*) at metaphase. Microtubules target the kinetochores of chromosomes prior to anaphase.

and target the kinetochores of the chromosomes for attachment (Fig. 7.11). The mitotic spindle checkpoint prevents anaphase until all of the chromatids are properly aligned at the metaphase plane. Note that a single unattached kinetochore is sufficient to arrest mitosis; it has been estimated that up to a quarter of the total kinetochore attachment time can be devoted to capturing the last kinetochore before anaphase commences.

The biochemical control of anaphase is summarized in Fig. 7.12. Anaphase is triggered by a cascade of inhibitory interactions initiated by the activation of APC/Cdc20 (see preceding chapter for discussion of this enzyme complex). The proteins called *cohesin* that glue sister chromatids together are degraded by the enzyme *separase* whose protease activity is inhibited by a protein called *securin*; the latter is a direct target of inhibition by APC/Cdc20. It is believed that the spindle checkpoint acts by inhibiting APC/Cdc20, but the details of how the checkpoint operates are still unclear. Extensive biochemical studies with budding yeast have demonstrated that a protein complex, called the *mitotic checkpoint complex* (MCC), forms at an unattached kinetochore. Components of the MCC include the proteins Mad2, Mad3/BubR1, Bub3, and Cdc20. Thus, the MCC competes against the APC for Cdc20, and this competition has been suggested as a possible mechanism of APC inhibition by the MCC. Another hypothesis emphasizes the role of some mechanical signalling via the lack of spindle tension due to an unattached kinetochore.

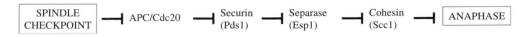

Fig. 7.12 The series of inhibitory interactions believed to occur from the 'spindle checkpoint' to the prevention of anaphase.

Starting with the assumption that the checkpoint signal originates from the last unattached kinetochore, Doncic *et al.* (2006) tested various models against two criteria for satisfactory checkpoint function, namely, (1) the ability of a single unattached kinetochore to keep APC/Cdc20 activity low throughout the nucleus, and (2) rapid increase in APC/Cdc20 activity after the last kinetochore is attached to microtubules. A satisfactory checkpoint model must take into account the physical dimensions of the cell and offer a mechanism for propagating the anaphase-inhibitory signal from a single unattached kinetochore to the entire chromosome assembly. In addition, experimental observations have shown that the anaphase signal must be activated within a short time after the final kinetochore is attached.

Among the models investigated by Doncic *et al.* (2006), the so-called 'emitted inhibition model' satisfies the criteria for satisfactory spindle-checkpoint function. This model assumes a molecular species that is activated only at the unattached kinetochore, and that once this molecule is activated, it can catalyze the inhibition of APC everywhere in the nucleus. At this time, the identity of the molecular species that is activated specifically at the unattached kinetochore is a matter of speculation. However, the analysis of Doncic *et al.* (2006) illustrates the impact that spatial and temporal constraints have on the design of the mitotic-spindle-checkpoint mechanism.

References

Aguda, B. D. and Tang, Y. (1999) 'The kinetic origins of the restriction point in the mammalian cell cycle', *Cell Proliferation* **32**, 321–335.

Aguda, B. D. (1999a) 'Instabilities in phosphorylation-dephosphorylation cascades and cell cycle checkpoints', *Oncogene* **18**, 2846–2851.

Aguda, B. D. (1999b) 'A quantitative analysis of the kinetics of the G2 DNA damage checkpoint system', *Proceedings of the National Academy of Sciences of the USA* **96**, 11352–11357.

Bartek, J. and Lukas, J. (2001) 'Mammalian G1- and S-phase checkpoints in response to DNA damage', *Current Opinion in Cell Biology* **13**, 738–747.

Doncic, A., Ben-Jacob, E. and Barkai, N. (2006) 'Evaluating putative mechanisms of the mitotic spindle checkpoint', *Proceedings of the National Academy of Sciences of the USA* **102**, 6332–6337.

Ekholm, S. V., Zickert, P., Reed, S. I., Zetterberg, A. (2001) 'Accumulation of cyclin E is not a prerequisite for passage through the restriction point,' *Molecular Cell Biology* **21**, 3256–3265.

Elledge, S. J. (1996) 'Cell cycle checkpoints: preventing an identity crisis', *Science* **274**, 1664–1672.

Pines, J. (1999) 'Four-dimensional control of the cell cycle,' *Nature Cell Biology* **1**, E73–E79.

Pardee, A. B. (1974) 'A Restriction Point for Control of Normal Animal Cell Proliferation', *Proceedings of the National Academy of Sciences of the USA* **71**, 1286–1290.

Exercises

1. Consider the differential equations for $[Y_1]$ and $[Y_2]$ described in the caption of Fig. 7.6. (For simplicity, use the symbols y_1 and y_2 for these concentrations, respectively.) Show that the non-zero steady-state concentrations y_1^s and y_2^s are explicitly given by

$$y_1^s = \frac{E_1 E_2 - c_1 c_2}{E_2 + c_1} \ , \ y_2^s = \frac{E_1 E_2 - c_1 c_2}{E_1 + c_2}, \ \text{where } c_i = \frac{k_{ir}}{k_{if}} \ (i = 1, 2) \text{ and } E_1 E_2 > c_1 c_2.$$

   Show also that transcritical bifurcation occurs at parameters that satisfy $E_1 E_2 = c_1 c_2$, and that the non-zero steady states are linearly stable.

2. As in Exercise 1, consider the system in Fig. 7.6(a) but now with Michaelis–Menten kinetics for v_{1f} and v_{2f}:

$$v_{1f} = \frac{k_{1f} y_2 x_1}{K_{M1} + x_1} \ , \ \ v_{2f} = \frac{k_{2f} y_1 x_2}{K_{M2} + x_2}.$$

   Does transcritical bifurcation occur? If it does, find the expression for the parameters where this type of bifurcation occurs, and determine the linear stability of the steady-state branches before and after they cross at the bifurcation point.

3. Using mass-action kinetics (as in the caption of Fig. 7.6), write down the differential equations for the network shown in Fig. 7.8 (Use the concentrations of Wee1, MPF, and Cdc25-P as independent variables.) Find the steady states of the network and determine their stability. Does a bifurcation occur in this system? If it does, determine the type of bifurcation and the expression for the parameters where this bifurcation occurs.

8
Cell death

Cells divide and differentiate to form various tissues and organs in the development of a multicellular organism. Along the way, some cells are destined or 'programmed' to die, perhaps according to an architectural blueprint of the adult. *Apoptosis* is the name given to this programmed cell death. Many of the essential genes, proteins, and molecular pathways regulating apoptosis are now known, and several mathematical models of these pathways have been proposed. This chapter will introduce the biological manifestations of apoptosis and gives an overview of the regulatory biochemical pathways. These pathways are classified as either intrinsic or extrinsic – the former being mediated by the mitochondria, and the latter by so-called death ligands. Kinetic models of these pathways are presented in this chapter. It is also increasingly recognized that apoptosis is the same mechanism that cells employ to eliminate cells with irreparably damaged DNA. Indeed, abnormalities in the regulation of apoptosis have been implicated in the origins of various human diseases.

8.1   Background on the biology of apoptosis

Cells die either by *necrosis* or by *apoptosis*; the former could be due to acute tissue injury leading to a messy death (Fig 8.1(a)), while the latter is a relatively 'clean' or orderly death (Fig 8.1(b)). In necrosis, the cell spills its contents and often elicits inflammatory responses. Apoptosis was first discovered in the context of development – for example, cells in a tadpole's tail disappear as the animal matures. Because certain cells are destined to die during development, apoptosis is often referred to as 'programmed cell death.' In the past few years, it has been increasingly recognized that apoptosis plays a key role in many human malignancies, including cancer, neurodegenerative disorders and autoimmune diseases. In certain tumors, cells with excessive DNA damage are supposed to die but do not do so because of faulty apoptosis machinery.

The word 'apoptosis' comes from the Greek word for 'falling off' or 'dropping off' as leaves falling from a tree. Its manifestations as observed under the microscope are: cell shrinkage, chromatin condensation, DNA fragmentation, membrane blebbing, collapse of the nucleus, disintegration of the cell into apoptotic bodies, and final lysis of apoptotic bodies.

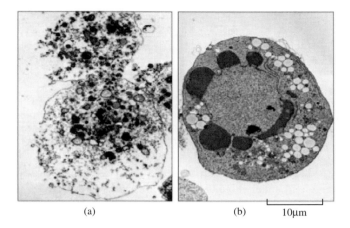

(a)   (b)   10μm

Fig. 8.1 Cell death by (a) necrosis and (b) apoptosis. (Pictures courtesy of Julia F. Burne).

8.2 Intrinsic and extrinsic caspase pathways

Central to the apoptotic molecular machinery is a group of proteases called *caspases* (acronym for *c*ysteine-containing *asp*artic-acid specific prote*ases*). Proteases are enzymes that catalyze the degradation of cellular proteins. Caspases induce cell death by cleaving cellular proteins such as nuclear lamins (building blocks of nuclear architecture), DNA repair enzymes, and various cytoskeletal proteins.

The general pathways for caspase-dependent apoptosis are conserved across many species. At least 14 different caspases have been identified so far (Philchenkov, 2004). Caspases are classified into two groups, namely, *initiator* caspases and *effector* (or *executioner*) caspases. Initiator caspases are the first to be activated upon a cell's exposure to death signals; the initiator caspases are said to be 'upstream' of effector caspases in a cascade of caspase activation. Certain thresholds of activity of effector caspases are required for the death sentence of cells. Caspase-8 and caspase-9 are examples of initiator caspases. Caspase-3, caspase-6, and caspase-7 are examples of executioner caspases. The inactive precursors of these caspases, called procaspases, are constitutively expressed by animal cells; their activation requires cleavage and association of the cleaved fragments to form the active site of the caspase (depicted in Figure 8.2). The active caspase enzyme is believed to be a tetramer of 2 large subunits and 2 small subunits (there is still controversy as to whether the dimer also possesses caspase activity). Procaspase-8 molecules when brought together have been observed to possess weak protease activities that may initiate a caspase-activation cascade.

Activation of effector caspases can be induced by pathways classified as 'extrinsic' (also called membrane receptor-mediated pathway) or 'intrinsic' (also called the mitochondria-mediated pathway). Caspase-8 and caspase-9 are initiator caspases of the extrinsic and intrinsic pathways, respectively. The pathways are depicted in Fig 8.3.

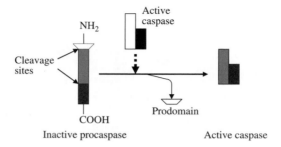

Fig. 8.2 Schematic diagram of the activation of a procaspase through cleavages that separate the prodomain, large and small subunits of the procaspase. An active caspase is shown to catalyze these cleavage reactions. Figure redrawn from Alberts *et al.* (2002).

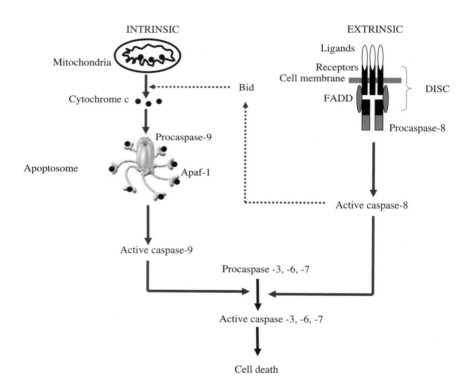

Fig. 8.3 A schematic representation of the extrinsic and intrinsic pathways of caspase-mediated apoptosis. DISC = death-inducing signalling complex. The apoptosome is composed of cytochrome c and a heptamer of Apaf-1. The picture of the apoptosome is from Beere (2005). Copyright 2005 American Society for Clinical Investigation.

In Fig 8.3, the extrinsic pathway begins with the formation of the DISC complex. DISC is the acronym for '*death-inducing signaling complex*'. It is a protein complex formed when an extracellular ligand (e.g. the homotrimeric Fas ligand) binds to membrane receptors, followed by the recruitment of a protein called FADD that acts as an 'adaptor' for procaspase-8 molecules. The colocalization of several procaspase-8 molecules at the DISC initiates activation of caspase-8. One of the models to be discussed in this chapter (in Section 8.4) focuses on the formation of DISC and activation of caspase-8.

The intrinsic pathway is initiated by a variety of intracellular stresses, including DNA damage, oxidative stress, and ischemia. These disturbances lead to the permeabilization of the outer mitochondrial membrane and release of cytochrome c, SMAC/Diablo, and other proapoptotic molecules. The complex labelled 'apoptosome' in Fig 8.3 is composed of a heptamer of Apaf-1 (apoptotic protease-activating factor) proteins, along with bound ATP and cytochrome c molecules. Procaspase-9 molecules are then 'recruited' to the apoptosome to generate active caspase-9. As shown by the arrow from caspase-8 to the arrow from mitochondria, the extrinsic pathway can also enhance the intrinsic pathway. The details in the regulation of the activities of the initiator and effector caspases will be discussed in the mathematical models below.

8.3   A bistable model for caspase-3 activation

The model proposed by Eissing *et al.* (2004) – hereafter referred to as the *Eissing model* – focuses on the extrinsic apoptotic pathway. The model considers a small number of steps that generates bistability. As discussed in Chapter 3, bistability is characterized by having two locally stable steady states that coexist for some fixed parameter values. The molecular processes involved in the Eissing model are shown in Figure 8.4 Important structural features of the network include the positive-feedback loop between caspase-8 and caspase-3, the mutual inhibition between caspase-3 and the protein IAP (inhibitor of apoptosis), and the inhibition of caspase-8 by the protein BAR. Note that the aforementioned mutual inhibition is also considered as a positive-feedback loop.

The individual reaction steps in the Eissing model are as follows (numbers above the arrows are labels matching those of Fig. 8.4; a negative label is for the reverse reaction):

$$C8^* + C3 \xrightarrow{1} C8^* + C3^*$$

$$C8 + C3^* \xrightarrow{2} C8^* + C3^*$$

$$C3^* + IAP \xleftrightarrow{3,-3} iC3^* \sim IAP$$

$$C3^* + IAP \xrightarrow{4} C3^*$$

$$C8^* \xrightarrow{5}$$

$$C3^* \xrightarrow{6}$$

$$iC3^* \sim IAP \xrightarrow{7}$$

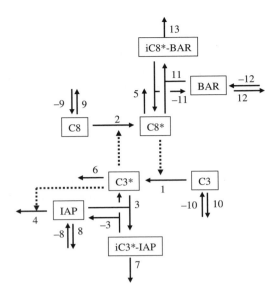

Fig. 8.4 The model of Eissing *et al.* (2004). A *dashed arrow* means catalysis of the step that the arrow points to. See text for details.

$$\text{IAP} \xrightarrow{8,-8}$$

$$\text{C8} \xrightarrow{9,-9}$$

$$\text{C3} \xrightarrow{10,-10}$$

$$\text{C8}^* + \text{BAR} \xrightarrow{11,-11} \text{iC8}^* \sim \text{BAR}$$

$$\text{BAR} \xrightarrow{12,-12}$$

$$\text{iC8}^* \sim \text{BAR} \xrightarrow{13}$$

When the right-hand side of the reaction arrow is blank (as in steps 5–10 and 12–13), it means that the protein is being degraded; steps (–8), (–9), (–10), and (–12) mean that constant rate of production of the protein is assumed. The rates of the individual reactions are given in Table 8.1. Each reversible reaction is written as two one-way reactions, each with its own rate expression. Note that mass-action kinetics is assumed for all rate expressions in Table 8.1

The dynamical equations for the Eissing model are listed in eqn 8.1 ([X] means 'concentration of X'; the expressions for the v_is are given in Table 8.1).

$$\frac{\mathrm{d}[C8]}{\mathrm{d}t} = v_{-9} - (v_9 + v_2)$$

$$\frac{\mathrm{d}[C8^*]}{\mathrm{d}t} = (v_2 + v_{-11}) - (v_5 + v_{11})$$

Table 8.1 Expressions for the reaction rates (v_i) and values of the kinetic parameters (k_j) in the model of Eissing *et al.* (2004).

Reaction rates, v_i	Parameter values, k_j
$v_1 = k_1 \,[\text{C8*}][\text{C3}]$	$k_1 = 5.8 \times 10^5 \text{ M}^{-1}\text{s}^{-1}$
$v_2 = k_2[\text{C3*}][\text{C8}]$	$k_2 = 10^5 \text{ M}^{-1}\text{s}^{-1}$
$v_3 = k_3[\text{C3*}][\text{IAP}]$	$k_3 = 5 \times 10^6 \text{ M}^{-1}\text{s}^{-1}$
$v_{-3} = k_{-3}[\text{iC3*} \sim \text{IAP}]$	$k_{-3} = 0.035 \text{ s}^{-1}$
$v_4 = k_4[\text{C3*}][\text{IAP}]$	$k_4 = 3 \times 10^6 \text{ M}^{-1}\text{s}^{-1}$
$v_5 = k_5[\text{C8*}]$	$k_5 = 5.8 \times 10^{-3} \text{ min}^{-1}$
$v_6 = k_6[\text{C3*}]$	$k_6 = 5.8 \times 10^{-3} \text{ min}^{-1}$
$v_7 = k_7[\text{iC3*} \sim \text{IAP}]$	$k_7 = 1.73 \times 10^{-2} \text{ min}^{-1}$
$v_8 = k_8[\text{IAP}]$	$k_8 = 1.16 \times 10^{-2} \text{ min}^{-1}$
$v_{-8} = k_{-8}$	$k_{-8} = 1.3 \times 10^{-11} \text{ M s}^{-1}$
$v_9 = k_9[\text{C8}]$	$k_9 = 3.9 \times 10^{-3} \text{ min}^{-1}$
$v_{-9} = k_{-9}$	$k_{-9} = 1.4 \times 10^{-11} \text{ M s}^{-1}$
$v_{10} = k_{10}[\text{C3}]$	$k_{10} = 3.9 \times 10^{-3} \text{ min}^{-1}$
$v_{-10} = k_{-10}$	$k_{-10} = 2.3 \times 10^{-12} \text{ M s}^{-1}$
$v_{11} = k_{11}[\text{C8*}][\text{BAR}]$	$k_{11} = 5 \times 10^6 \text{ M}^{-1}\text{s}^{-1}$
$v_{-11} = k_{-11}[\text{iC8*} \sim \text{BAR}]$	$k_{-11} = 0.035 \text{ s}^{-1}$
$v_{12} = k_{12}[\text{BAR}]$	$k_{12} = 10^{-3} \text{ min}^{-1}$
$v_{-12} = k_{-12}$	$k_{-12} = 1.1 \times 10^{-12} \text{ Ms}^{-1}$
$v_{13} = k_{13}[\text{iC8*} \sim \text{BAR}]$	$k_{13} = 60 \text{ min}^{-1}$

$$\frac{\text{d}[C3]}{\text{d}t} = v_{-10} - (v_1 + v_{10})$$

$$\frac{\text{d}[C3^*]}{\text{d}t} = (v_1 + v_{-3}) - (v_3 + v_6)$$

$$\frac{\text{d}[\text{IAP}]}{\text{d}t} = (v_{-3} + v_{-8}) - (v_3 + v_8 + v_4) \tag{8.1}$$

$$\frac{\text{d}[iC3^* \sim \text{IAP}]}{\text{d}t} = v_3 - (v_{-3} + v_7)$$

$$\frac{\text{d}[\text{BAR}]}{\text{d}t} = (v_{-11} + v_{-12}) - (v_{11} + v_{12})$$

$$\frac{\text{d}[iC8^* \sim \text{BAR}]}{\text{d}t} = v_{11} - (v_{-11} + v_{13}).$$

Equations (8.1) are solved to generate the plots in Fig. 8.5 that show the system switching from a state with low caspase-3 ('alive' state) activity to a state with high activity ('dead' state). A steady-state bifurcation diagram can also be

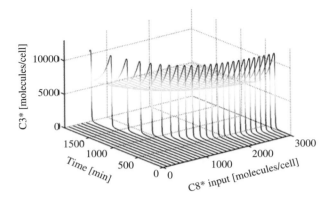

Fig. 8.5 Dynamics of active caspase-3 (C3*) for various initial values of active caspase-8 (C8*). Equations (8.1) and parameters in Table 8.1 are used in the computer simulations. Figure is reproduced with permission (Eissing *et al.*, 2004).

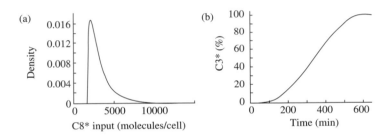

Fig. 8.6 The distribution of cells (density) according to input C8* shown in (a) gives rise to the slow C3* kinetics of the cell population shown in (b) when the dynamics of C3* activation (determined as in Fig. 8.5) is averaged over the population. Figures are reproduced with permission (Eissing *et al.*, 2004).

constructed to show the existence of bistability for some sets of parameter values (see Exercise 1).

Figure 8.5 shows that when the initial C8* value is less than a certain threshold, the system attains the state 'alive' (i.e. [C3*] remains close to zero); above this threshold, and after a pronounced time lag (which increases as initial C8* decreases), the system rapidly switches to the 'dead' state (high [C3*] state). These time lags and the fast kinetics of caspase-3 activation in single cells have been observed in laboratory experiments. Such fast kinetics is not observed in populations of cells that are known to be undergoing apoptosis. It could take 30 min to a few hours for caspase activation at the population level, while it could take less than 15 min at the single-cell level. These observations are accounted for by the Eissing model, as illustrated in Fig. 8.6. The population kinetics (average [C3*] versus time) shown in Fig. 8.6(b) can be generated by the cell number density profile shown in Fig. 8.6(a) as a function of initial number of C8* molecules per cell. This non-uniformity of initial number of C8* molecules can be rationalized by differences in expression levels of death receptors and/or of inhibitor proteins such as BAR in individual cells.

8.4   DISC formation and caspase-8 activation

Figure 8.3 shows the formation of DISC (death-inducing signalling complex) as a pre-requisite for the activation of the initiator caspase-8. Lai and Jackson (2004) analyzed a detailed mechanistic model that involves binding of extracellular 'death ligands' to membrane receptors (also called 'death receptors'), followed by a sequence of protein–protein interactions, and culminating in the recruitment of procaspase-8 molecules. Procaspase-8 molecules are known to possess weak protease activity and their co-localization at the DISC is believed to be sufficient for initiating a cascade of caspase activation. Families of death ligands and receptors have been identified. One of the most studied and understood ligand–receptor combinations is FasL/Fas. FasL is a homotrimeric ligand that binds to the Fas membrane receptor. Fas, also called CD95 or Apo1, is a member of the TNF (tumor necrosis factor) gene superfamily. The aim of the Lai–Jackson model is to understand the consequences of the 'multivalency' of FasL (i.e. this ligand's ability to bind up to three receptors at a time) on the kinetics of DISC formation and subsequent activation of caspase-8. The details of the model (with a few modifications of the original Lai–Jackson model) are given in Fig. 8.7 and Table 8.2.

To write the dynamical equation for a species shown in Fig. 8.7, one adds all the rates of incoming steps and subtracts the rates of all outgoing steps. The full system of equations is given below (the symbol for the concentration of a molecular species is identical to its symbol shown in Fig. 8.7 and Table 8.2):

$$\frac{dL}{dt} = v_{-1} - v_1$$

$$\frac{dR}{dt} = (v_{-1} + v_{-2} + v_{-3}) - (v_1 + v_2 + v_3)$$

$$\frac{dC_1}{dt} = (v_1 + v_{-2}) - (v_{-1} + v_2)$$

$$\frac{dC_2}{dt} = (v_2 + v_{-3} + v_{-4}) - (v_{-2} + v_3 + v_4)$$

$$\frac{dC_3}{dt} = (v_3 + v_{-5}) - (v_{-3} + v_5)$$

$$\frac{dD_{21}}{dt} = (v_4 + v_{-8} + v_{-6}) - (v_{-4} + v_8 + v_6)$$

$$\frac{dD_{22}}{dt} = (v_6 + v_{-10} + v_{14}) - (v_{-6} + v_{10})$$

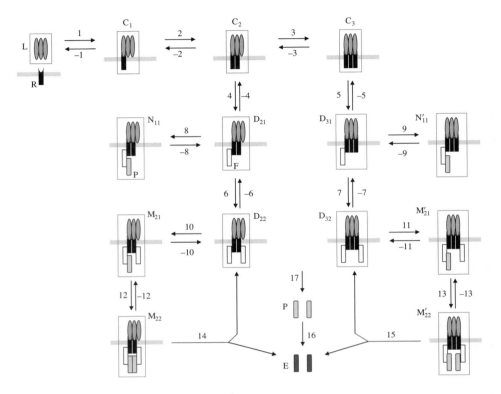

Fig. 8.7 The Lai–Jackson model. *Horizontal grey bar* represents the cell membrane. L = ligand, R = membrane receptor, F = FADD, P = procaspase-8, E = executioner caspase.

$$\frac{\mathrm{d}D_{31}}{\mathrm{d}t} = (v_5 + v_{-7} + v_{-9}) - (v_{-5} + v_7 + v_9)$$

$$\frac{\mathrm{d}D_{32}}{\mathrm{d}t} = (v_7 + v_{-11} + v_{15}) - (v_{-7} + v_{11})$$

$$\frac{\mathrm{d}N_{11}}{\mathrm{d}t} = v_8 - v_{-8}$$

$$\frac{\mathrm{d}M_{21}}{\mathrm{d}t} = (v_{10} + v_{-12}) - (v_{-10} + v_{12})$$

$$\frac{\mathrm{d}M_{22}}{\mathrm{d}t} = v_{12} - (v_{-12} + v_{14})$$

$$\frac{\mathrm{d}N'_{11}}{\mathrm{d}t} = v_9 - v_{-9}$$

$$\frac{\mathrm{d}M'_{21}}{\mathrm{d}t} = (v_{11} + v_{-13}) - (v_{-11} + v_{13})$$

(8.2)

Table 8.2 Parameters of the Lai–Jackson model. The symbols for the concentrations of the various molecular species are identical to their symbols shown in Fig. 8.7

Reaction rates, v_i	Correspondence of parameter values, k_j, (left column of this table) and parameter symbols (in bold) used by Lai and Jackson (2004)
$v_1 = k_1 \text{LR}$	$k_1 = \mathbf{3k}_f$
$v_{-1} = k_{-1}\text{C}_1$	$k_{-1} = \mathbf{k}_r$
$v_2 = k_2\text{C}_1\text{R}$	$k_2 = \mathbf{2k}_x$
$v_{-2} = k_{-2}\text{C}_2$	$k_{-2} = \mathbf{2k}_{-x}$
$v_3 = k_3\text{C}_2\text{R}$	$k_3 = \mathbf{k}_x$
$v_{-3} = k_{-3}\text{C}_3$	$k_{-3} = \mathbf{3k}_{-x}$
$v_4 = k_4\text{C}_2\text{F}$	$k_4 = \mathbf{2k}_f$
$v_{-4} = k_{-4}\text{D}_{21}$	$k_{-4} = k_{-1}$
$v_5 = k_5\text{C}_3\text{F}$	$k_5 = k_4$
$v_{-5} = k_{-5}\text{D}_{31}$	$k_{-5} = k_{-4}$
$v_6 = k_6\text{D}_{21}\text{F}$	$k_6 = k_3$
$v_{-6} = k_{-6}\text{D}_{22}$	$k_{-6} = k_{-2}$
$v_7 = k_7\text{D}_{31}\text{F}$	$k_7 = k_6$
$v_{-7} = k_{-7}\text{D}_{32}$	$k_{-7} = k_{-6}$
$v_8 = k_8\text{D}_{21}\text{P}$	$k_8 = \mathbf{k}_p$
$v_{-8} = k_{-8}\text{N}_{11}$	$k_{-8} = \mathbf{k}_{-p}$
$v_9 = k_9\text{D}_{31}\text{P}$	$k_9 = k_8$
$v_{-9} = k_{-9}\text{N'}_{11}$	$k_{-9} = k_{-8}$
$v_{10} = k_{10}\text{D}_{22}\text{P}$	$k_{10} = k_8$
$v_{-10} = k_{-10}\text{M}_{21}$	$k_{-10} = k_{-8}$
$v_{11} = k_{11}\text{D}_{32}\text{P}$	$k_{11} = k_8$
$v_{-11} = k_{-11}\text{M'}_{21}$	$k_{-11} = k_{-8}$
$v_{12} = k_{12}\text{M}_{21}\text{P}$	$k_{12} = \mathbf{k}_c$
$v_{-12} = k_{-12}\text{M}_{22}$	$k_{-12} = \mathbf{k}_{-c}$
$v_{13} = k_{13}\text{M'}_{21}\text{P}$	$k_{13} = k_{12}$
$v_{-13} = k_{-13}\text{M'}_{22}$	$k_{-13} = k_{-12}$
$v_{14} = k_{14}\text{M}_{22}$	$k_{14} = \mathbf{k}_a$
$v_{15} = k_{15}\text{M'}_{22}$	$k_{15} = k_{14}$
$v_{16} = k_{16}\text{P}^2$	$k_{16} = \bar{\mathbf{k}}_a$
$v_{17} = k_{17}$	k_{17}

$$\frac{dM_{22}'}{dt} = v_{13} - (v_{-13} + v_{15})$$

$$\frac{dF}{dt} = (v_{-4} + v_{-6} + v_{-5} + v_{-7}) - (v_4 + v_6 + v_5 + v_7)$$

$$\frac{dP}{dt} = (v_{17} + v_{-8} + v_{-10} + v_{-12} + v_{-9} + v_{-11} + v_{-13})$$
$$- (2v_{16} + v_8 + v_{10} + v_{12} + v_9 + v_{11} + v_{13})$$

$$\frac{dE}{dt} = 2v_{14} + 2v_{15} + 2v_{16}.$$

As can be deduced from Fig. 8.7, the system of eqn 8.2 is subject to several conservation constraints: the total concentration of all ligand-containing species is a constant (L_0), the total concentration of all receptor-containing species is a constant (R_T), and the total concentration of all FADD-containing species is a constant (F_T).

Strictly speaking, reversible transitions between D_{21} and D_{31}, between D_{22} and D_{32}, between N_{11} and M_{21}, and between N'_{11} and M'_{21}, should be included in a more complete model; however, the discussion below will be restricted to the steps shown in Fig. 8.7 and eqn 8.2.

Consider the L-R-C_1-C_2-C_3 subsystem with only the following rates: v_1, v_{-1}, v_2, v_{-2}, v_3, and v_{-3} (see Table 8.2). As depicted in Fig. 8.7, only clustered receptors (at least two) can propagate the 'death signal' through their ability to recruit the adaptor protein FADD; the latter has protein structural domains called 'death domains' that bind to complementary death domains present in the cytoplasmic portion of the Fas receptor. Since ligand–receptor interactions are fast, the L-R-C_1-C_2-C_3 subsystem is assumed to reach steady state quickly compared to the events after receptor clustering. The steady states of L, R, C_1, C_2, and C_3 can be determined by equating to zero all the right-hand sides of the corresponding equations in eqn 8.2, and taking into account the conservation constraints on the total ligand (L_0) and total receptor (R_T) concentrations (see Exercise 2).

The total number of clustered pairs of Fas receptors, C_{XT}, is used to monitor the intensity of the death signal (Lai and Jackson refers to C_{XT} as the total number of receptor cross-links); it is defined as follows:

$$C_{XT} = C_{2,\text{eq}} + 2C_{3,\text{eq}}, \tag{8.3}$$

where $C_{2,\text{eq}}$ and $C_{3,\text{eq}}$ are the steady-state (equilibrium) values of C_2 and C_3, respectively. Plots of C_{XT} versus L_0 (for a fixed R_T) show how the death signal varies as a function of total ligand concentration. As Fig. 8.8 shows, for a given R_T (i.e. fixed κ in the figure), a plot of C_{XT} has a maximum at a unique value $L_{0,\text{max}}$; the maximum C_{XT} value increases (while the corresponding $L_{0,\text{max}}$ decreases) as R_T (or κ) increases. The plots in Fig. 8.8 also show that if there is a threshold value of C_{XT} (e.g. shown as $C_{XT,\text{min}}$ in the figure) below which no death signal is propagated, then this threshold requires that there is a corresponding minimum value κ_{min} (equivalently, $R_{T,\text{min}}$ for death signal propagation. For $\kappa > \kappa_{\text{min}}$, only ligand concentrations within a certain

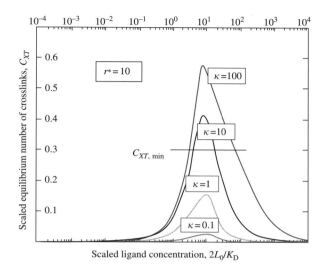

Fig. 8.8 Number of equilibrium crosslinks eqn 8.2 as a function of scaled total ligand concentration. $r^* = R_T/K_D$, R_T = total receptor concentration, L_0 = total ligand concentration, $K_D = k_r/k_f$, $\kappa = K_x R_T$, $K_x = k_x/k_{-x}$, (see Table 8.2 for meaning of k_r, k_f, k_x, k_{-x}). Figure reproduced with permission from Lai and Jackson (2004).

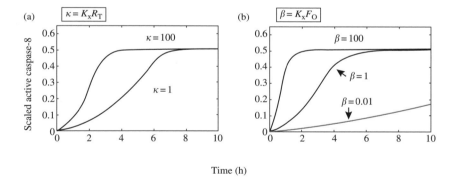

Fig. 8.9 (a) The effect of FAS trimerization (represented by the parameter κ) on the activation of caspase-8. (b) The effect of FADD binding disruption (represented by the parameter β) on the activation of caspase-8. $K_x = k_x/k_{-x}$, R_T = total receptor concentration, F_o = total FADD concentration. Figures reproduced with permission from Lai and Jackson (2004).

range will successfully propagate the death-signal; the smaller $(\kappa - \kappa_{\min})$ is, the narrower this range will be. The assumption that there exists a $C_{XT,\min}$ implies that the apoptosis threshold can be controlled at a stage upstream of the caspase cascade.

Figure 8.9(a) shows the dynamics of activation of caspase-8 for two values of κ (proportional to the total receptor concentration, R_T). Figure 8.9(b) shows the effect of changing FADD concentration (proportional to β) on caspase-8 activation. These plots demonstrate the significance of the κ and β parameters in the activation of the initiator caspase-8.

8.5 Combined intrinsic and extrinsic apoptosis pathways

Fussenegger *et al.* (2000) proposed a detailed mechanistic model involving both intrinsic and extrinsic pathways for the activation of executioner caspases. The detailed network is shown in Fig. 8.10. All of the steps shown in this figure are reversible except the five steps tagged with asterisks (*) and the step for cytochrome c release (labelled r_C). The extrinsic pathway (top half of the network) includes binding of the ligand (L) to the membrane receptor (R), a step for clustering of the death domains of the receptor ($\tilde{R}L \rightarrow RL$), sequential binding of two FADD (F) molecules, sequential recruitment of two procaspase-8 molecules (c_{8z}) to the $RL.F_2$ complex, and the colocalized c_{8z} proteins cleaving each other to generate active caspase-8 (c_{8a}).

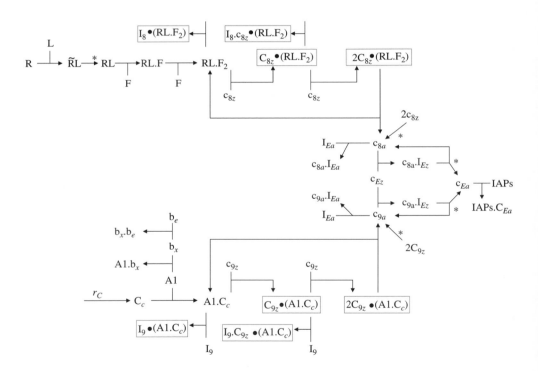

Fig. 8.10 The model of Fussenegger *et al.* (2000) for the activation of caspases. R = membrane receptor, L = ligand, F = FADD, c_{kz} = procaspase-k (k = 8, 9, E), I_k = 'decoy' proteins inhibiting c_{kz} (k = 8, 9), c_{ja} = active caspase-j (j = 8, 9, E), I_{Ea} = 'decoy' proteins inhibiting c_{8a} or c_{9a}, IAPs = inhibitors of apoptosis, A1 = Apaf-1, C_c = cytochrome c, b_x = anti-apoptotic protein of the Bcl-2 family, b_e = proapoptotic protein of the Bcl-2 family, r_C = rate of release of cytochrome c from mitochondria.

The model includes 'decoy' proteins (I_8) – such as proteins called FLIPs – that possess death-effector domains similar to FADD and could compete with procaspase-8 for binding sites on FADD. The inhibition of active caspase-8 or caspase-9 by other 'decoy' molecules (I_{Ea}) is also a part of the model. Also included is the possibility, though of small rate, for two procaspase-8 molecules to activate each other without the mediation of the DISC.

For the intrinsic pathway, the model assumes the following rate of cytochrome c release

$$r_C = \alpha_{CE} v(c_{Ea}, b_x), \tag{8.4}$$

where α_{CE} is the specific rate of cytochrome c efflux due to active executioner caspases (Fussenegger *et al.* also considered the contributions of nutritional factors but these are not included in eqn 8.4 for simplicity). The factor $v(c_{Ea}, b_x)$ is a function of the active executioner caspase (c_{Ea}) and antiapoptotic members of the Bcl-2 family (represented as b_x in Fig. 8.10); $v(c_{Ea}, b_x)$ is assumed to be proportional to the ratio $[c_{Ea}]/[b_x]$ and to be within the range $0 \leq v(c_{Ea}, b_x) \leq 1$. It is further assumed that there exists a threshold value (ε) for the ratio $[c_{Ea}]/[b_x]$ such that cytochrome c is released if $[c_{Ea}]/[b_x] > \varepsilon$ (in this case $v(c_{Ea}, b_x) = 1$) and cytochrome c is not released if $[c_{Ea}]/[b_x] \leq \varepsilon$ (in this case $v(c_{Ea}, b_x) = 0$). It is important to note that the dependence of r_C on c_{Ea} implicitly assumes a positive feedback between active executioner caspases and the process of cytochrome c release. The regulation of cytochrome c release is not quite understood at this time, and this part of the model should be considered tentative. The interactions among antiapoptotic and proapoptotic Bcl-2 family members (b_x and b_e, respectively), as would be reflected in the ratio $[b_e]/[b_x]$, are thought to be crucial in the decision to die or survive. In the model, b_x competes against cytochrome c for Apaf-1 binding sites; on the other hand, proapoptotic b_e exerts its effect by binding to b_x so that the latter is prevented from binding to Apaf-1.

Cytochrome c binds to Apaf-1 ($A1$) to form a complex ($A1.C_c$) that recruits procaspase-9 (c_{9Z}) molecules and subsequently generate active caspase-9 (c_{9a}). There are 'decoy' proteins (I_9) – such as the protein called ARC – that compete against procaspase-9 for binding sites on ($A1.C_c$). As with caspase-8, decoy molecules (I_{Ea}) also inhibit caspase-9 (c_{9a}). It is also assumed that two procaspase-9 molecules can activate each other without involving the apoptosome. As depicted in Fig. 8.10, the pathway from $RL.F_2$ to c_{8a} has an identical structure to the pathway from $A1.C_c$ to c_{9a} and, correspondingly, the kinetic equations are of the same forms. The active initiator caspases (c_{8a} and c_{9a}) cleave and activate the executioner caspases (c_{Ea}). A family of proteins called IAPs (inhibitors of apoptosis) serves as the final line of defense for the cell; IAPs directly regulate the activity of executioner caspases by binding to them.

The protein Bid links the extrinsic to the intrinsic apoptosis pathway (see Fig 8.3) but is not included in the model of Fussenegger *et al.* However, even if only the extrinsic pathway is initially active, the model of Fussenegger *et al.* predicts that the intrinsic pathway will eventually be activated as a result of the positive feedback from the executioner caspases to the process of cytochrome c release (see expression for the rate

Table 8.3 Effects of therapies predicted by the model of Fussenegger *et al.* (2000). As examples, overexpression of Bcl-2 alone *cannot* ('−') decrease executioner caspase activation, but overexpression of IAPs alone *can* ('+'); and the combination of overexpression of FLIPs *and* mutation of p53 *can* ('+') also decrease executioner caspase activation. Note that no single therapy, except overexpression of IAPs, inhibits activation of executioner caspases when both intrinsic and extrinsic apoptosis pathways are active; therapy combinations shown in this table are required. (Table from Fussenegger *et al.*, 2000).

	Overexpression						Disruption or mutation	
	Bcl-2	Bcl-x$_L$	Bax	ARC	FLIPs	IAPs	FADD	p53
Bcl-2	−	−	−	−	+	+	+	−
Bcl-x$_L$	−	−	−	−	+	+	+	−
Bax	−	−	−	−	−	+	−	−
ARC	−	−	−	−	+	+	+	−
FLIPs	+	+	−	+	−	+	−	+
IAPs	+	+	+	+	+	+	+	+
FADD	+	+	−	+	−	+	−	+
p53	−	−	−	−	−	+	+	−

of cytochrome *c* release, r_C, in eqn 8.4). From the network diagram of Fig. 8.3, one can give answers to the question on what genes or proteins can be perturbed to block the activation of executioner caspases. Table 8.3, taken from Fussenegger *et al.* (2000), gives examples of combined 'therapies' (i.e. perturbations such as overexpression or mutation of genes) that can block activation of executioner caspases. The column on p53 is based on the information (not shown in the network diagram of Fig 8.10) that p53 induces expression of proapoptotic Bax, and that p53 antagonizes the antiapoptotic proteins Bcl-x$_L$ and Bcl-2. Among the listed genes or proteins in Table 8.3, the only successful single therapy is the overexpression of IAPs (note that this is the only diagonal entry with a + sign). Overexpression of any of the antiapoptotic Bcl-2 family members (Bcl-2 and Bcl-x$_L$) cannot block executioner caspase activation simply because the extrinsic pathway is active. Similarly, overexpression of FLIPs and disruption of FADD binding (both proteins are involved in inhibiting the extrinsic pathway) cannot block executioner caspase activation because the intrinsic pathway is active. The table also shows that overexpression of FLIPs and mutation of p53 allows blocking of executioner caspase activation; this is because overexpression of FLIPs can block the extrinsic pathway while inactivation of p53 permits the antiapoptotic Bcl-2 proteins to block the intrinsic pathway.

8.6 Summary and future modelling

All the models discussed in this chapter focus on the activation of proteolytic enzymes called *caspases* whose targets include proteins of the cytoskeleton, the nuclear lamina, enzymes involved in fragmenting DNA, and others that are directly or indirectly associated with the cell-morphological manifestations of apoptosis. The irreversibility of this cell-death process is no doubt reinforced by the cascading nature, with concomitant amplification, of the activation of caspases. There are two caspase cascades considered in the models, namely, one that is a membrane receptor-mediated (termed the *extrinsic* pathway) and the other involving mitochondria (termed the *intrinsic* pathway). The Eissing model focuses on the extrinsic pathway represented by caspase-8 (initiator) and caspase-3 (executioner). The magnifying character of this caspase cascade is implemented by a positive-feedback loop between these two caspases (in other words, caspase-8 activates caspase-3 that, in turn, generates more active molecules of caspase-8). A significant proposal based on the Eissing model is that the transition between a living cell (low caspase-3) and a dying cell (high caspase-3) is governed by a bistable switch – this is a prediction that needs experimental validation at the single-cell level. The Eissing model also demonstrates the role of caspase inhibitors (such as IAP and BAR in the model) in generating a time lag prior to the fast-switching kinetics of caspase-3 activation.

The Lai–Jackson model is concerned with the consequences of the 'multivalent' (trimer) character of the death ligand FasL on the formation of the death-inducing signalling complex (DISC) and on the activation of initiator caspase-8. One of the interesting predictions of this model is that, given a fixed number of membrane death receptors, the number of DISCs that are formed increases and then decreases as the concentration of death ligands increases (see Fig. 8.8); this implies that if a minimum number of DISCs is required to trigger apoptosis, then ligand concentrations must be within a given range.

A model of the intrinsic pathway was published (after this chapter was written) by Legewie *et al.* (2006). This interesting work highlights the role of positive-feedback loops in the network in generating bistability, including the contribution to bistability that is mediated by IAPs (inhibitors of apoptosis).

A model that combines the extrinsic and intrinsic pathways is proposed by Fussenegger and coworkers. This model considers neither the trimeric character of FasL, nor the positive feedback loop between caspase-3 and caspase-8; it does, however, implicitly assume a positive feedback between caspase-3 and the process of cytochrome *c* release from mitochondria. The model accounts for interactions at the outer mitochondrial membrane involving antiapoptotic and proapoptotic members of the Bcl-2 family. The current literature on the subject of Bcl-2 proteins seems to claim that the ratio of the numbers of antiapoptotic and proapoptotic member proteins is a crucial factor in the decision to undergo apoptosis. The feasibility of this claim can be investigated by further modelling and analysis. In addition, future modelling of the combined extrinsic and intrinsic caspase pathways should also explore the role of the protein Bid (see Fig. 3.3) that links the extrinsic pathway to the intrinsic pathway. The molecular network regulating apoptosis is far from complete, as recent investigations strongly suggest that there are apoptosis-inducing molecular pathways that do not involve the caspases; thus, new mechanistic models of apoptosis are expected to be created in the near future.

References

Alberts, B., Johnson, A., Lewis, J., Raff, M., Roberts, K., and Walter, P. (2002) *Molecular biology of the cell*, 4th edn. Garland Publishing, New York.

Beere, H. M. (2005) 'Death versus survival: functional interaction between the apoptotic and stress-inducible heat shock protein pathways,' *Journal of Clinical Investigation* **115**, 2633–2639.

Eissing, T., Conzelmann, H., Gilles, E. D., Allgower, F., Bullinger, E., Scheurich, P. (2004) 'Bistability analyses of a caspase activation model for receptor-induced apoptosis', *Journal of Biological Chemistry.* **279**, 36892–36897.

Fussenegger, M., Bailey, J. E., and Varner, J. (2000) 'A mathematical model of caspase function in apoptosis,' *Nature Biotechnology* **18**, 768–774.

Lai, R. and Jackson, T. L. (2004) 'A mathematical model of receptor-mediated apoptosis: dying to know why FASL is a trimer,' *Mathematical Biosciences and Engineering* **1**, 325–338.

Legewie, S., Bluthgen, N., and Herzel, H. (2006) 'Mathematical modeling identifies inhibitors of apoptosis as mediators of positive feedback and bistability,' *PLoS Computational Biology* **2(9)**, e120.

Philchenkov, A. (2004) 'Caspases: potential targets for regulating cell death,' *Journal of Cellular and Molecular Medicine* **8**, 432–444.

Exercises

1. Determine the steady states of the Eissing model eqn 8.1. Plot the steady states of C3* as a function of k_1, and determine the linear stability of these steady states. Verify the claim of the authors that if the protein BAR is not included in the model (i.e. exclude steps 11, –11, 12, –12, 13, and –13), the system exhibits an 'alive' steady state that is only stable when k_1 is below $\sim 3.2 \times 10^3$ M^{-1}s^{-1}. Use the parameter values given in Table 8.1.

2. For the Lai–Jackson model summarized in Fig. 8.7, Table 8.2, and eqns 8.2, show that the steady state is unique and linearly stable for a fixed set of parameters.

3. Using the steady-state assumption for the L-R-C_1-C_2-C_3 subsystem of the Lai–Jackson model, determine the ligand concentration (L_0) that corresponds to the maximum of the C_{XT} curve for a given κ (see Fig. 8.8). If there exists a $C_{XT,min}$ below which the death signal is not propagated, determine the minimum total receptor concentration $R_{T,min}$ that allows propagation; and for $R_T > R_{T,min}$, determine the range of ligand concentrations that allow propagation of the death signal.

4. Using Fig 8.7 (Lai–Jackson model), and ignoring steps 16 and 17, show that the steady-state rate of production of active caspase-8 (that is, $(dE/dt)_{eq}$) has the form

$$\left(\frac{dE}{dt}\right)_{eq} = L_T(\alpha R_{eq}^2 + \beta R_{eq}^3),$$

where L_T is the total ligand concentration in the system.

5. Explain all the entries (+ or –) in Table 8.3 using the network diagram of the Fussenegger model shown in Fig 8.10.

9
Cell differentiation

The development of a multicellular organism from a single fertilized egg cell to various cells with specialized functions, in tissues and organs, is a most fascinating biological process. How does this cell give rise to an intricately structured adult organism with an almost deterministic precision? In embryogenesis, cells divide, migrate, and acquire different programs of gene expression determining whether they will be part of the endoderm, the ectoderm, or the mesoderm (see Fig. 9.1). Following this initial stage of cell *determination* is a maturation process called *differentiation* by which cells acquire recognizable phenotypes and functions. These phenotypes are the results of the expression of specific proteins; for example, muscle cells produce the myosin protein needed for muscle contraction, and red blood cells synthesize hemoglobin for oxygen transport in the blood.

Somatic cells of various differentiation lineages generally possess identical genomes but differ in their patterns of gene expression – this is why models of cell differentiation focus on gene-transcription regulatory networks. Questions on whether there is some sort of a genetic 'blueprint' for the construction of the mature organism and how this genetic information unfolds during development are profound questions that are very difficult to answer at this time. However, it is believed that answers will come from studies of epigenetic mechanisms modifying chromatin structure and, consequently, the gene-transcription machinery. Epigenetic mechanisms do not alter the sequence of DNA bases but, instead, change chromatin components through covalent modifications (e.g. DNA methylation, histone acetylation, histone phosphorylation).

In this chapter, the kinetic modelling of T helper (Th) lymphocyte differentiation is discussed. This is one of the better understood examples of cell differentiation, both at the genetic and molecular levels. T lymphocytes originate from the differentiation of hematopoietic ('blood-making') stem cells. The hematopoietic system with its various cell lineages is briefly introduced in Section 9.1. Essential details of the molecular control of Th differentiation and a dynamical model – proposed by Yates, Callard and Stark (YCS) – are described in Section 9.2. As will be illustrated for single cells in Section 9.3, differentiation can be controlled by two transcription factors exhibiting intrinsic bistable behavior. In Section 9.4, the YCS model is used to study the interaction between extracellular and intracellular signalling in a population of Th cells. Lastly, abstract models of high-dimensional switches (with more than two differentiation outcomes possible) are presented in Section 9.5.

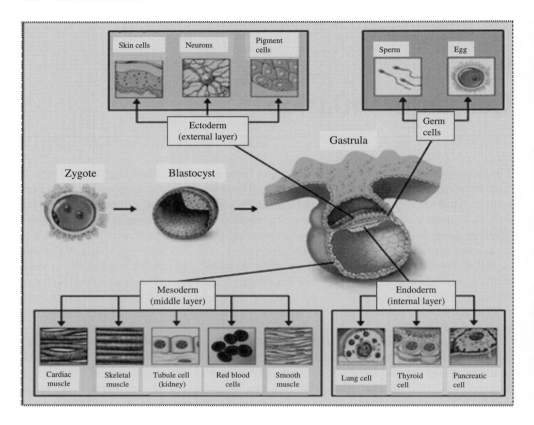

Fig. 9.1 Human embryogenesis. After the fertilization of an egg to create the zygote, rapid mitotic divisions with no significant growth occur to form a fluid-filled sphere of cells called the blastocyst. During the process of gastrulation, cells specialize and migrate to various positions in the embryo. The inner cell mass gives rise to germ cells (sperm or eggs) and to cells of the three germ layers (ectoderm, mesoderm, and endoderm). Reproduced with permission from the Office of Science Policy, the National Institutes of Health, USA.

9.1 Cell differentiation in the hematopoietic system

The lineages from the hematopoietic stem cell to the specialized cells in the blood and lymph are summarized in Fig. 9.2. Further differentiation of T lymphocytes is the subject of the next three sections of this chapter.

Lymphocytes are white blood cells that play important roles in the immune system, the body's defense against pathogens. T cells and B cells are two major types of lymphocytes. T cells are so-called because the immature T cells migrate to the thymus gland where they mature. B cells are so-called because they mature in the bone marrow (in mammals). B cells produce antibodies against pathogens (such as bacteria), while T cells are involved in autoimmunity (the ability to eliminate body cells that are

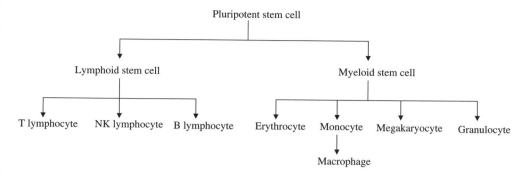

Fig. 9.2 The pluripotent hematopoietic stem cell gives rise to multipotent stem cells of the lymphoid and myeloid lines. Myeloid stem cells differentiate into erythrocytes (red blood cells), megakaryoctes (involved in blood clotting), granulocytes, monocytes and macrophages (the last three cells are involved in the immune system, particularly in digesting or engulfing pathogens and dead cells). Lymphoid stem cells give rise to T, NK (natural killer), and B lymphocytes. See text for more details.

infected by viruses or have become tumorigenic). Th lymphocytes represent a subtype of T cells and are identified by the presence of surface antigens (proteins) called CD4 (hence the cells are referred to as $CD4^+$). Other subtypes of T cells include cytotoxic T cells ($CD8^+$) and regulatory T cells ($CD4^+CD25^+$, also known as suppressor T cells). CD8 and CD25 are other types of surface antigens. Cytotoxic T cells target and destroy infected body cells. Regulatory T cells suppress the activation of the immune system.

Th cells are by far the most numerous of the T cells in a healthy individual. When Th cells get activated they proliferate and secrete cytokines that help regulate the autoimmune response. In fact, $CD4^+$ T cells are one of the known targets of the HIV virus; abnormally low levels of these Th cells result in AIDS (acquired immunodeficiency syndrome). The modelling of Th lymphocyte differentiation to cell types called Th1 and Th2 and the acquisition of cytokine memory are discussed in the next two sections.

9.2   Modelling the differentiation of Th lymphocytes

T-cell activation refers to the stimulation of growth and proliferation of T-cells via a mechanism involving the interaction of T-cell receptors with antigens on the surfaces of antigen-presenting cells (APCs). After an initial antigenic stimulation, Th lymphocytes differentiate into either of two distinct types called Th1 and Th2. These cells are distinguished by the repertoire of cytokines they produce. Th1 cells make IFNγ and lymphotoxin needed for combating intracellular pathogens (this is the so-called cellular immune response that, if abnormal, is associated with inflammatory and autoimmune diseases). Th2 cells produce the cytokines IL-4, IL-5, and IL-13 that activate B cells to

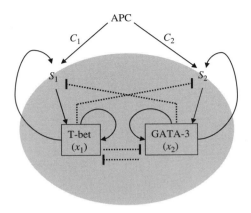

Fig. 9.3 Network diagram corresponding to the Yates–Callard–Stark model of Th lymphocyte differentiation driven by the cross-antagonism between two autoactivating transcription factors, T-bet (x_1) and GATA-3 (x_2). Cytokine signals S_1 and S_2 are dependent on the intracellular levels of x_1 and x_2, respectively, as well as non-helper T cell signals C_i ($i = 1, 2$) such as those from antigen-presenting cells (APCs).

produce antibodies against extracellular pathogens (this is called the humoral immune response that, if abnormal, is associated with allergies and asthma). Whether a precursor Th cell (symbolized by Th0) becomes Th1 or Th2 depends on 'polarizing' signals. Examples of Th1-polarizing signalling molecules are IL-12 (interleukin-12) and IFNγ (interferon-γ). An example of a Th2-polarizing signalling molecule is IL-4. An objective for modelling this system is to understand the phenomenon of 'cytokine memory' exhibited by Th1 and Th2 cells – that is, after an initial antigenic stimulation by APCs, cells with cytokine memory no longer require additional polarizing signals to express the cytokines characteristic of their differentiated state.

The YCS model of Th differentiation (Yates *et al.* 2004) is based on the interactions of two transcription factors, T-bet and GATA-3. These interactions and the molecular pathways regulating them are summarized in Fig. 9.3. High protein levels of T-bet or GATA-3 correspond to the Th1 phenotype or the Th2 phenotype, respectively.

Two important features of the network shown in Fig. 9.3 are (1) the positive-feedback loops of T-bet and GATA-3, and (2) the cross-antagonism between these factors. As shown next, these network features generate a binary (either/or) decision during the differentiation process.

The autoactivation of each transcription factor is carried out by two positive feedback loops (Fig. 9.3). One is at the transcriptional level in which the protein promotes the expression of its corresponding gene by binding to the gene's promoter region. The other positive feedback involves an autocrine loop – that is, a cell expresses a protein that is secreted into the extracellular medium, and this protein then binds to a membrane receptor of the same cell type; the ensuing activation of the receptor activates a pathway that eventually induces the expression of the same protein. For Th1 cells, a major autocrine loop involves IFNγ, IFNγR, STAT1 signalling, and T-bet (IFNγR = IFNγ receptor, STAT = signal transducer and activator of transcription). These Th1 autocrine loops are represented by the loop connecting S_1 and x_1 (T-bet) in Fig. 9.3. For Th2 cells, a major autocrine loop involves IL-4, IL-4R, STAT6 signalling, and GATA-3 (IL-4R = IL-4 receptor). Th2 autocrine loops are represented by the loop connecting S_2 and x_2 (GATA-3).

The cross-antagonism between T-bet and GATA-3 also occurs at two levels as depicted in Fig. 9.3 (dashed lines). One level involves transcriptional repression of one factor by the other, while the other level involves pathways from the transcription factors to inhibition of membrane receptors.

Th1-polarizing and Th2-polarizing cytokines as well as other cytokines that stimulate T-bet and GATA-3 production are symbolized by S_1 and S_2, respectively. The intracellular concentrations of T-bet and GATA-3 are symbolized by x_1 and x_2. In the YCS model, the dynamics of x_1 and x_2 are described by:

$$\frac{dx_1}{dt} = -\mu x_1 + \left(\alpha_1 \frac{x_1^n}{\kappa_1^n + x_1^n} + \sigma_1 \frac{S_1}{\rho_1 + S_1} \right) \times \frac{1}{(1 + x_2/\gamma_2)} + \beta_1$$
$$\frac{dx_2}{dt} = -\mu x_2 + \left(\alpha_2 \frac{x_2^n}{\kappa_2^n + x_2^n} + \sigma_2 \frac{S_2}{\rho_2 + S_2} \right) \times \frac{1}{(1 + x_1/\gamma_1)} + \beta_2. \tag{9.1}$$

The first term on the right-hand side of each equation in eqn 9.1 represents the rate of protein degradation (assumed to be first order in the protein concentration, and with identical decay rate constant μ for both proteins). The last term β_i is the constant basal rate of protein synthesis. The autoactivation rate of protein x_i is represented by the term

$$\alpha_1 \frac{x_1^n}{\kappa_1^n + x_1^n},$$

where n is the Hill exponent that tunes the sharpness of the activation switch. The contribution of external signalling to the rate of growth in x_i is given by the term

$$\sigma_i \frac{S_i}{\rho_i + S_i}.$$

The cross-inhibition between x_1 and x_2 occurs at both the autoactivation level and external (membrane) signalling level, and is represented by the cross-inhibition factors

$$\frac{1}{(1 + x_i/\gamma_i)}.$$

The parameter γ_i represents the value of x_i at which the rate of production of $x_j, i \neq j$, (due to the combined autoactivation and external signalling) is halved.

9.3 Cytokine memory in single cells

Considering the dynamics of each transcription factor independently (eqn 9.2), one can show that each factor is intrinsically bistable (Fig. 9.4).

$$\frac{dx_i}{dt} = -\mu x_i + \alpha_i \frac{x_i^n}{\kappa_i^n + x_i^n} + \sigma_i \frac{S_i}{\rho_i + S_i} + \beta_i \quad (i = 1, 2). \tag{9.2}$$

Figure 9.4(a) predicts that for Th0 precursor cells in the states represented by branch A of the curve, increasing S_2 leads to a threshold level θ_2 above which a transition to a high steady-state level of x_2 occurs (the C branch shown in Fig. 9.4(a)). Once on the C branch, small fluctuations in the signal S_2 will not return the system to the Th0 state, unless S_2 is close to θ_1 where even a small decrease in S_2 below this value can return the system to the Th0 state. Thus, in a biological setting where cells are stimulated to grow and proliferate, Fig. 9.4(a) shows that as long as external signalling is above θ_1, cells that have acquired the high-x_2 phenotype will maintain this phenotype – in other words, these cells have acquired *cytokine memory*. However, continuous external signalling above the θ_1 threshold is required to maintain progeny cells at high x_2 levels; otherwise, the cells return to the precursor Th0 phenotype.

In contrast to the reversibility between the differentiated state and the Th0 state, Fig 9.4(b) illustrates the commitment of cells to a differentiated state once it is reached. The steady-state curve is shifted to the left when the parameter α_2 is increased (or alternatively decreasing the decay rate μ) and the lower threshold value of θ_1 disappears, thus making impossible a return to the Th0 state if the system is in the high-x_2 branch of the steady-state curve. Experimentally, this commitment or irreversibility in differentiation is observed with Th lymphocytes after 4 or 5 cell divisions.

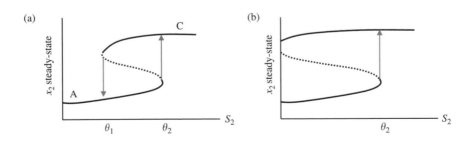

Fig. 9.4 Steady states of x_2 as a function of the level of cytokine signalling, S_2 for two different values of the parameter α_2. Figure adapted from Yates *et al.* (2004).

9.4 Population of differentiating Th lymphocytes

9.4.1 Equation for population density Φ

The YCS model can be used to simulate the dynamics of a population of Th lymphocytes in an extracellular medium, itself affected by secretions from individual cells. Because of the bistable character of each cell and the cross-antagonism between the two differentiated states, one would like to investigate whether there are certain conditions when the population becomes homogeneous in just one type of cell or whether there are conditions that allow mixtures of differentiated cells.

In the original study by the authors of the YCS model, the cell population is assumed to be spatially homogeneous but with time-dependent distribution of T-bet and GATA-3 levels. A population density $\Phi(x_1, x_2, t)$ is defined such that the number of cells (per unit volume) with levels of T-bet and GATA-3 in the ranges $[x_1, x_1 + \delta x_1]$ and $[x_2, x_2 + \delta x_2]$, respectively, at time t is $\Phi(x_1, x_2, t)\delta x_1 \delta x_2$. To incorporate in the model the coupling between extracellular and intracellular signalling, each cytokine signals $S_i(i = 1, 2)$ is assumed to depend on the total level of the corresponding intracellular transcription factor x_i (see the integral involving x_i in the equation below) and on signals from non-T helper cell sources (the $C_i(t)$ term in the equation below):

$$S_i = \frac{C_i(t) + \int x_i \Phi \mathrm{d}x_1 \mathrm{d}x_2}{\int \Phi \mathrm{d}x_1 \mathrm{d}x_2}, \tag{9.3}$$

where $0 \leq x_i \leq \infty$. Note that S_i is normalized by the total Th cell population size (i.e. S_i is a measure of the cytokine signal *per cell*). Equations 9.1 and 9.3 imply that each cell senses the population average of x_i for the autocrine loop (positive feedback) between S_i and the transcription factor x_i (Fig. 9.3).

If g is the rate at which cells divide (number of divisions per unit time), the time evolution of Φ is given by the following conservation of mass equation:

$$\frac{\partial \Phi}{\partial t} + \frac{\partial (f_1 \Phi)}{\partial x_1} + \frac{\partial (f_2 \Phi)}{\partial x_2} = g\Phi, \tag{9.4}$$

where $f_1(x_1, x_2) = \mathrm{d}x_1/\mathrm{d}t$, and $f_2(x_1, x_2) = \mathrm{d}x_2/\mathrm{d}t$ are given by eqns 9.1 and 9.3. Experimental observations suggest the initial value of $g = 2$ day^{-1}. At the end of this section, the method of solving eqn 9.4 with appropriate initial and boundary conditions is discussed.

The cross-suppression of the two differentiated states (giving rise to mutually exclusive population of Th1 or Th2 cells, but not mixed) and the switching between states (T0, Th1, or Th2) are demonstrated in the simulations shown in Fig. 9.5. Starting with a population of Th1 cells (i.e. 100% of the population with high T-bet), Th2-polarizing cytokine level C_2/N_0 (where N_0 is the initial total number of cells and C_2 is the extrinsic Th2-polarizing cytokine such as IL-4) is increased from 0.1 to 200 units per cell. Note that all the cells revert back to the Th0 state, and are kept in this precursor state for a wide range of C_2/N_0 values. A sudden switch to the Th2 state occurs at $C_2/N_0 \sim 60\text{--}70$ units per cell.

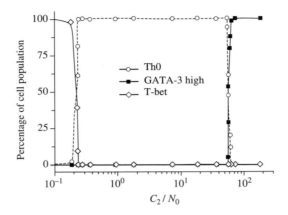

Fig. 9.5 Th1 populations of cells are subjected to various doses of Th2-polarizing cytokine, C_2/N_0, and then the fractions of cells in the Th0, Th1, and Th2 states are determined at day 6. Figure reproduced with permission from Yates *et al.* (2004). Copyright 2004 Elsevier Ltd.

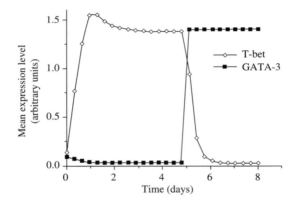

Fig. 9.6 Th0 cells are stimulated with a Th1-polarizing cytokine; although this cytokine is assumed to decay with a half-life of 8 h, the Th1 state is maintained (see first 5 days). GATA-3 is ectopically expressed in all cells on day 5. Figure reproduced with permission from Yates *et al.* (2004). Copyright 2004 Elsevier Ltd.

Experiments have been reported in which ectopic expression of T-bet in committed Th2 cells lead to a switch to the Th1 state without first going through a reversion to the Th0 state (Yates *et al.* 2004). The converse has also been shown experimentally – that is, ectopic expression of GATA-3 in Th1 cells switch them to the Th2 state. The YCS model exhibits the same phenomenon as demonstrated in the simulations shown in Fig. 9.6.

■ 9.4.2 Determining the population density Φ

In order to solve eqn 9.4, initial and boundary conditions have to be imposed. Consider the special case where it is not possible to have cells with x_1 or x_2 vanishing at any time, that is

$$\Phi(0, x_2, t) = \Phi(x_1, 0, t) = 0, \tag{9.5}$$

and, in addition, that there are no cells with x_1 or x_2 very large (let X represent this large positive number), that is

$$\Phi(X, x_2, t) = \Phi(x_1, X, t) = 0. \tag{9.6}$$

Let the initial value of Φ be

$$\Phi(x_1, x_2, 0) = \Phi_0(x_1, x_2). \tag{9.7}$$

From eqns 9.1 and 9.5, one can see that

$$f_1(0, x_2) > 0, \quad f_2(x_1, 0) > 0, \tag{9.8}$$

and from eqns 9.1 and 9.6

$$f_1(X, x_2) < 0, \quad f_2(x_1, X) < 0, \tag{9.9}$$

provided $\mu X > \alpha_i + \sigma_i + \beta_i$ for $i = 1, 2$. Let the domain R be defined by

$$R = \{(x_1, x_2) | 0 < x_1 < X \text{ and } 0 < x_2 < X\}. \tag{9.10}$$

If S_i $(i = 1, 2)$ were known, eqn 9.4 with boundary and initial conditions 9.5–9.7 can be solved by the method of characteristics. Note that eqns 9.8 and 9.9 ensure that characteristics starting at $x_i = 0$ or $x_i = X (i = 1, 2)$ enter the domain R defined above.

$S_i(t)$ in eqn 9.3 is determined using a sequence of successive approximations, $S_i^n(t)$. Suppose $S_i^n(t)$ has already been computed; eqn 9.4 is then solved by replacing S_i $(i = 1, 2)$ in eqn 9.1 with S_i^n. Denote the solution Φ by φ_i^n, and define

$$S_i^{n+1}(t) = \frac{C_i(t) + \int\limits_R x_i \varphi_i^n \mathrm{d}x_1 \mathrm{d}x_2}{\int\limits_R \varphi_i^n \mathrm{d}x_1 \mathrm{d}x_2}. \tag{9.11}$$

To complete the definition of the sequence S_i^n, set

$$S_i^0(t) = \frac{C_i(t) + \int\limits_R x_i \Phi_0 \mathrm{d}x_1 \mathrm{d}x_2}{\int\limits_R \Phi_0 \mathrm{d}x_1 \mathrm{d}x_2}. \tag{9.12}$$

With $S_i = \lim\limits_{n \to \infty} S_i^n$ and $\Phi = \lim\limits_{n \to \infty} \varphi_i^n$, one thus arrives at the solution Φ of eqn 9.4 with the imposed boundary and initial conditions.

9.5 High-dimensional switches in cellular differentiation

The Th differentiation described in the previous sections has only two possible outcomes, Th1 or Th2 states. This binary decision is implemented by a network with cross-antagonism between two autoactivating transcription factors (T-bet and GATA-3); the autoactivating property is important in reinforcing a decision once it is made. There are many other examples of transcription factors controlling binary decisions during the development of multicellular organisms. There are also quite a number of examples in which a cell may differentiate into any of three or more lineages, with differentiation steps that cannot be reduced to sequences of binary decisions. For example, hermaphrodite germline cells of the nematode *C. elegans* can differentiate into a somatic cell, sperm, or oocyte (egg). Cinquin and Demongeot (2005) reviewed evidence of other multi-outcome differentiation systems, and studied simple models composed of cross-antagonizing and autocatalytic factors. A high-dimensional differentiation switch is compared with the corresponding cascade of binary switches in Fig. 9.7.

The binary cascade starting from the left of Fig 9.7 involves three pairs of cross-antagonizing factors. The Cinquin–Demongeot (CD) model of 'one-shot' differentiation shown on the right of Fig. 9.7 involves four cross-antagonizing factors. In both binary and CD models, the progenitor cells express all the switch elements, albeit in small quantities, before each differentiation step; for example, in Fig 9.7, the progenitor cell at the left of the binary cascade starts with low-level expressions of factors G_1 and G_2. Upon receiving differentiation cues, either G_1 or G_2 increases, giving rise to either

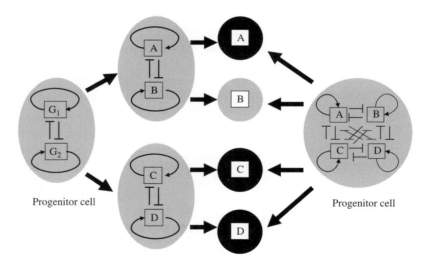

Fig. 9.7 Four differentiated cells (A, B, C, D) depicted by the four small circles, from a cascade of binary switch elements (from left progenitor cell) or from a progenitor cell with four coupled switch elements (from right progenitor cell).

of the cells shown in the second column of Fig. 9.7. Each of these progeny cells has two competing transcription factors that will then make further binary decisions.

One of the Cinquin–Demongeot (2005) models is discussed here to illustrate potential differentiation scenarios not exhibited by binary-decision models. Let there be n transcription factors (also called switch elements) whose concentrations are denoted by $x_i(i = 1, \ldots, n)$. In the CD model, the dynamics of the interactions among these elements are described by the following equations:

$$\frac{\mathrm{d}x_i}{\mathrm{d}t} = -x_i + \frac{\sigma x_i^c}{1 + \sum_{i=1}^{n} x_i^c} + \alpha \quad (i = 1, \ldots, n). \tag{9.13}$$

For simplicity, every switch element is assumed to have a first-order degradation rate with rate coefficient of 1. The second term on the right-hand side of eqn 9.13) is the rate of synthesis of each element that is positively regulated by itself (autoactivation) and inhibited by all other elements (as represented by the sum in the denominator). The third term represents a constant basal rate of expression of each element. The 'co-operativity' parameter c tunes the sharpness of the autoactivation switch and the parameter σ is related to the strength of gene expression.

The model assumes that the steady states of the system 9.13 represent the differentiated states. For simplicity, let $c = 2$ and $\alpha = 0$, and investigate the steady states as the strength of gene expression σ is increased. Setting the right-hand side of eqn 9.13 to zero gives

$$\sigma x_{i,s}^2 = x_{i,s}\left(1 + \sum_{i=1}^{n} x_{i,s}^2\right), \tag{9.14}$$

where $x_{i,s}$ is the steady-state value of x_i. Let $\mathbf{x}_s = (x_{1,s}, x_{2,s}, \ldots x_{n,s})$ be the vector of steady states. Clearly, $(0, 0, \ldots, 0)$ is always a steady state. Because of the symmetry of the equations, all positive $x_{i,s}$ are equal to each other and the values for $m \ (\geq 1)$ positive elements must satisfy

$$\sigma x_{i,s} = 1 + m x_{i,s}^2. \tag{9.15}$$

Thus, the parameter σ must satisfy a condition, namely, $\sigma^2 \geq 4m$, for $\mathbf{x}_s$ to have at least m elements that are positive. As a specific example, consider a system with two elements $(n = 2)$. For $0 \leq \sigma^2 < 4$, the only possible steady state is the origin. For $4 < \sigma^2 < 8$ the system has new steady states of the form $(0, p)$ and $(p, 0)$, in addition to the origin; from eqn 9.15 one can see that p has two possible values, namely, $p_+ = (\sigma + \sqrt{\sigma^2 - 4})/2$ or $p_- = (\sigma - \sqrt{\sigma^2 - 4})/2$. Thus, when $4 < \sigma^2 < 8$, a system with two elements has a total of 5 steady states, namely, $(0, 0)$, $(0, p_-)$, $(0, p_+)$, $(p_-, 0)$, and $(p_+, 0)$. These steady states and state-space trajectories are shown in Fig. 9.8. The unstable steady states in the boundary (unfilled circles) are saddle points whose stable manifolds (the trajectories that approach these steady states in infinite time) separate the basins of attractions of the three stable steady states (black circles). As x_1 and x_2 increase, the basin of attraction of the steady state at the origin shrinks and the probability of finding the system in either the high-x_1 or high-x_2 steady state increases. This can be interpreted as a cell-differentiation scenario

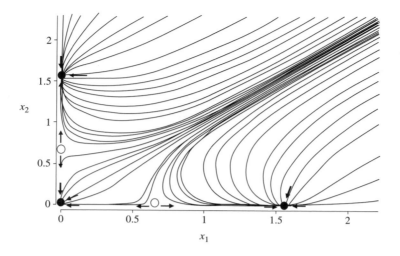

Fig. 9.8 Phase portrait of the system described in eqn 9.13 for two switch elements ($n = 2$), $\alpha = 0$, and $\sigma = 2.2$. The small circles show the locations of the steady states; as indicated by the directions of the trajectories, the *black circles* are locally stable and the *unfilled circles* are unstable.

in which a precursor cell (identified with the steady state at the origin) initiates its differentiation by increasing the levels of x_1 and x_2 ('coexpression'), thereby decreasing the likelihood of the cell keeping the precursor phenotype and increasing the chance that the cell attains either the high-x_1 phenotype or the high-x_2 phenotype.

For $\sigma^2 \geq 8$, the two-element system has additional steady states of the form (p, p) where $p_\pm = (\sigma \pm \sqrt{\sigma^2 - 8})/4$. Thus, the model predicts two new cell phenotypes associated with the system steady states (p_+, p_+) and (p_-, p_-); whether or not these phenotypes can be observed biologically depends on the stability of the steady states and the size of their basins of attraction. The phase portrait shown in Fig. 9.8 can be generalized to any number of switch elements; thus, for the system depicted on the right of Fig. 9.7, four cross-antagonizing switch elements can give rise to four differentiated states depending on the initial conditions (in biological terms, these would be the 'differentiation cues' that favor one steady state over another). Each differentiated state corresponds to a high level of one of the four switch elements and vanishing levels of the others.

9.6 Summary

The Yates–Callard–Stark (YCS) model of T helper (Th) lymphocyte differentiation is based on the cross-antagonism between two transcription factors – namely, T-bet and GATA-3 – whose mutually exclusive dominance determines whether the cell attains a Th1 or a Th2 phenotype, respectively. These transcription factors control the expression of specific genes for different cytokines that characterize phenotypes. Once a direction of differentiation is chosen, it is reinforced by autocrine loops and by the autoactivating property of the transcription factor. Due to these positive-feedback loops, the YCS model predicts bistability and the existence of cytokine signalling thresholds for switching the system from the precursor cell phenotype (Th0) to one of the differentiated phenotypes.

The coupling between intracellular and extracellular signalling in a population of Th cells was also investigated using the YCS model. The model predicts plasticity (reversibility) among the Th0, Th1, and Th1 states, as illustrated in computer simulations where the population is subjected to increasing cytokine signalling (polarized towards one differentiated state) or ectopic expression of one of the transcription factors. However, commitment to one phenotype (i.e. irreversibility) is not excluded by the YCS model as demonstrated in Fig. 9.4(b). Non-Th polarizing cytokine signals – represented by the term $C_i(t)$ in eqn 9.3 – link the YCS model to 'differentiation cues'; particularly in models that encompass spatial patterning in development, these differentiation cues have to be explicitly considered. Current models of development adapt the idea that concentrations of molecules called morphogens are sensed by cells as differentiation cues. Morphogens can be proteins such as transcription factors or ligands whose concentration gradients elicit differentiation of cells into at least two distinct phenotypes. The morphogen hypothesis is reviewed recently by Ashe and Briscoe (2006).

Lastly, the Cinquin–Demongeot (CD) model of cellular differentiation with more than two transcription factors (switch elements) generating more than two phenotype outcomes was discussed. As in the YCS model, every pair of switch elements are cross-antagonistic and each element is autoactivating. The CD model predicts that increasing a parameter for the rate of gene expression of the switch elements leads to a rapidly increasing number of steady states (equivalently, to an increasing number of differentiated states).

References

Ashe, H. L. and Briscoe, J. (2006) 'The interpretation of morphogen gradients', *Development* **133**, 385–394.

Cinquin, O. and Demongeot, J. (2005) 'High-dimensional switches and the modeling of cellular differentiation', *Journal of Theoretical Biology* **233**, 391–411.

Mariani, L., Lohning, M., Radbruch, A., and Hofer, T. (2004) 'Transcriptional control networks of cell differentiation: insights from helper T lymphocytes', *Progress in Biophysics & Molecular Biology* **86**, 45–76.

Yates, A., Callard, R., and Stark J. (2004) 'Combining cytokine signalling with T-bet and GATA-3 regulation in TH1 and Th2 differentiation: a model for cellular decision-making', *Journal of Theoretical Biology* **231**, 181–196.

Exercises

1. Mariani *et al.* (2004) also modelled Th differentiation but, in contrast to the YCS model, the mRNAs of the transcription factors are considered explicitly. For example, the GATA-3 dynamics (independent of T-bet) is modelled by the following

differential equations (using the symbols in their paper):

$$\frac{dG}{dt} = k_T R - k_G G$$

$$\frac{dR}{dt} = v_B + v_{\max,S}\frac{S}{K_S + S}$$

$$+ v_{\max,G}\left(\frac{G}{K_G + G}\right)^2 - k_R R,$$

where G is the protein concentration of GATA-3, R is the concentration of its mRNA, S is the concentration of a signalling protein (Stat6), and v_B is the basal rate of transcription. Show that the system above exhibits bistability similar to that of Fig. 9.4(a) of the YCS model.

2. With the boundary and initial conditions given by eqns 9.5–9.7, compute the solution of eqn 9.4 in the following cases:

   (a) $C_1(t) = c$, $C_2(t) = 0$, $\Phi_0(x_1, x_2) = \delta x_1 x_2 (X - x_1)(X - x_2)$
   (b) $C_1(t) = C_2(t) = c$, $\Phi_0(x_1, x_2) = \delta x_1 x_2 (X - x_1)(X - x_2)$

   where c and δ are positive constants. Assume that $\alpha_1 > \alpha_2, \sigma_1 = \sigma_2$, and $\beta_1 = \beta_2$.

3. Consider eqn 9.13 with $c = 2, n > m$, and $4m < \sigma^2 < 4(m+1)$ where m is the number of switch elements with positive steady states. Show that:

   (a) in addition to the steady state at the origin, the vector of steady states $\boldsymbol{x}_s$ has k elements that are zero and ℓ elements that are equal to $p \neq 0$ where

   $$\sigma p = 1 + \ell p^2, 1 \leq \ell \leq m.$$

   (b) the steady states having non-zero elements $p = (\sigma - \sqrt{\sigma^2 - 4\ell})/2$ are unstable and those with $p = (\sigma + \sqrt{\sigma^2 - 4\ell})/2$ are stable (here, stability means that the real parts of the eigenvalues of the Jacobian matrix evaluated at the steady state are all negative; instability means that at least one eigenvalue has a positive real part.)

4. Extend the result of the previous problem to the case $c > 2, n = 2$. Prove that, depending on σ, the only steady states are of the form $(0, 0)$, $(0, p)$, $(p, 0)$, and (p, p) where p has two possible values, p_1 and p_2, with $p_1 < p_2$. Also show that the steady state with $p = p_2$ is stable, and that with $p = p_1$ is unstable.

10
Cell aging and renewal

Cellular aging, commonly referred to as cellular *senescence*, is thought to be a 'programmed' cell fate in a similar sense that apoptosis is somehow programmed to happen to certain cells during development. A specific kind of aging called replicative senescence is characterized by a cell's permanent exit from the cell cycle after undergoing a certain number of divisions. One of the known causes of replicative senescence is telomere shortening after each division. Telomeres are the ends of linear eukaryotic chromosomes. A model involving replicative senescence of cells lining the walls of blood vessels is discussed in this chapter. Both a probabilistic model tracking individual cells and a deterministic model of the cell population dynamics are illustrated.

The lifespan of multicellular organisms would be significantly shorter were it not for innate processes of tissue renewal or regeneration. Everyday, for an average person, about 1.5 g of skin cells are shed and replaced, and about 200 billion red blood cells are replenished. These cells are generated by stem cells that divide and produce progenitors of many types of cells, in particular those replacing dead cells in tissues. Recently, there has been considerable interest in embryonic stem cells because of their ability to generate all of the tissues of an adult human. A model of stem-cell dynamics is discussed in this chapter. The model investigates the parameters affecting the maintenance of the stem-cell pool in their growth environments. Both probabilistic and deterministic simulations of the model are presented.

10.1 Cellular senescence and telomeres

Biologists make a distinction between cellular *quiescence* and *senescence*; both processes involve exit from the cell cycle, but quiescence is a temporary exit (i.e. cells re-enter the cell cycle upon sufficient mitogenic stimulation) while senescence is a permanent one. Besides cell-cycle arrest, other biomarkers of senescence include larger and more diverse morphology (see Fig. 10.1), distinct changes in gene expression, and telomere shortening.

Cellular senescence is thought to play a cancer-preventing role because exit from the cell cycle decreases the probability of acquiring DNA damage. Besides telomere attrition and DNA damage, other documented causes of cellular senescence are increased oxidative stress, lack of nutrients or growth factors, and improper cell–cell contacts.

In the early 1960s, Hayflick and Moorhead observed that human embryonic cells in culture can only divide 50 times, on average, before they enter into a permanently

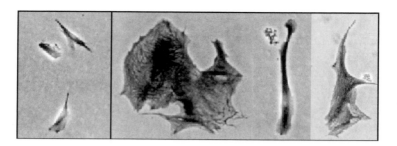

Fig. 10.1 Normal human fibroblasts are shown in the leftmost picture. Senescent human fibroblasts are shown in the other three pictures on the right. Reproduced with permission from João Pedro de Magalhães (http://www.senescence.info/).

non-dividing state. The number 50 is referred to as the *Hayflick limit* for these human cells. There is a correlation between the longevity of the organism and the Hayflick limit. For mice, the limit is about 15, while for the Galapagos tortoise (that could live over a century) the limit is about 110.

The existence of a Hayflick limit is explained in terms of telomere shortening. It is as if the telomeres provide a counting mechanism instructing a cell to stop dividing when the telomeres become shorter than some critical lengths. In the case of germ cells (egg and sperm), telomere lengths are maintained by the activity of an enzyme called telomerase that catalyzes telomere lengthening. Established cell lines used in *in-vitro* experiments are immortalized cells that express telomerase. These cells divide indefinitely as long as they are cultured in suitable conditions. An example of an immortalized cell line is the *HeLa* cell line that originates from cancer cells of *H*enrietta *La*cks who died of cervical cancer in 1951.

10.2   Models of tissue aging and maintenance

10.2.1   The probabilistic model of Op den Buijs *et al.*

Senescence of individual cells as a cause of aging of the organism is a controversial idea at this time, although some supporting evidence is available; for example, individuals with progeroid syndromes (characterized by premature aging) such as Werner syndrome have cells that exhibit replicative senescence at a much earlier time than normal. A recent model that deals with aging, not at the organismal level but at the tissue level, is discussed in this section. This is the model of Op den Buijs *et al.* (2004) involving endothelial cells – the cells that form the monolayer lining of the inner wall of blood vessels. It is a model that uses telomere shortening as the indicator of cell and tissue aging.

The endothelium is modeled by Op den Buijs *et al.* (henceforth called the *Buijs model*) as a square monolayer of endothelial cells (ECs). Computer simulations were carried out involving 500 ECs covering an area of $3.5 \times 10^5 \, \mu m^2$ (the cell radius is 15 μm). The model monitors the changes of the EC monolayer for 65 years, with a discrete time step of 1 year. ECs can be damaged and die, and dead cells could be replaced either by division of surrounding cells or by homing of endothelial progenitor cells (EPCs) – according to certain probabilities based on experimental evidence. The cells that die are chosen at random for each time step, and if the dead cells are to be replaced by cell division, non-senescent neighboring cells were randomly chosen to divide. It was assumed that the rates of EPC production from the bone marrow and EPC homing are such that there is a steady-state number of EPCs in the blood. EPCs that have homed are considered to convert instantly to fully differentiated EC.

A parameter that corresponds to telomere length, L, decreases by ΔL after each cell division. The value of ΔL is not fixed but is taken to be a geometric random variable; mean values of 50 to 250 base pairs (bp) per division were used in the simulations. A 20-year old individual is the starting point of the Buijs model, with initial telomere lengths that are distributed normally with mean 8 kbp and standard deviation of 2 kbp. The critical telomere length at which cells become senescent is taken to be 2 kbp (based on some measurements of human fibroblasts). The stem cells are assumed to be non-senescent for 65 years.

Figure 10.2 shows a set of computer simulations of the Buijs model. The pictures show the structures of the endothelium at age 65 if there is no EPC homing (left picture) and if there is 5% EPC homing (right picture; this percentage means that

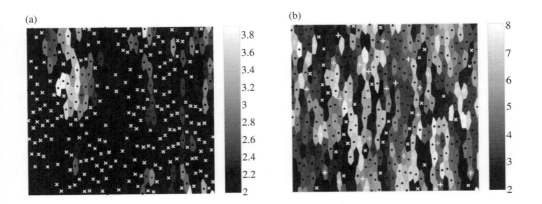

Fig. 10.2 The cellular profile of the endothelium wall of a 65-year old person according to simulations using the model of Op den Buijs *et al.* (2004). Homing of endothelial progenitor cells (EPCs) is absent in (a), and 5% EPC homing rate is assumed in (b). Grey levels on the right of each picture represent lengths (in thousands of DNA base pairs) of the proliferation-restricting telomeres per cell. Cells marked × are senescent; cells marked with + originated from EPC homing, and *black areas* are denuded of endothelial cells. Figures reproduced with permission from Op den Buijs *et al.* (2004).

up to 5% of endothelial cells are due to EPC homing). The simulation was carried out under abnormally increased telomere attrition rate ($\Delta L = 250$ bp/cell division) due to increased oxidative stress. It is seen that without the contribution of EPC homing, the majority of the cells are senescent and large denuded areas are present. In contrast, as shown on the right picture, 5% EPC homing is able to maintain the endothelium wall (with just a few denuded areas and few senescent cells). Note that the Buijs model should not be construed to mean that telomere-dependent aging of ECs is the only or primary cause of vascular disease; the modelling merely suggests that telomere-dependent senescence is a contributing factor.

10.2.2   A continuum model

The small number of ECs considered in the simulations described above of the Buijs model does not allow one to generalize the results to larger areas of the vasculature, where hundreds of thousands of ECs are involved. It is of interest to know how endothelium maintenance and damaging factors play out in a large population of ECs, because this information will then permit one to relate the aging of the organism to the aging of the EC wall. Wang, Aguda & Friedman (2007) developed a continuum mathematical model (henceforth called the *WAF model*) of the endothelium describing how densities of ECs of different telomere lengths evolve in time. This model is discussed below.

The dynamical variable $m_i(x, y, t)$ represents the number density (cells per unit area) of cells of generation i (this index counts the number of mitotic divisions until senescence). The number of cells of generation i at time t in the area bounded by x and $x + \mathrm{d}x$, and y and $y + \mathrm{d}y$ is equal to $m_i(x, y, t)\mathrm{d}x\mathrm{d}y$. For viable cells the index i ranges from 0 (for EPCs that just homed on the endothelium) to N (for senescent cells). The value taken for N is 50, the human Hayflick limit. For dead cells, $i = N+1$. To account for the space created when dead cells are cleared, a variable called 'hole density', $h(x, y, t)$, is defined to be identical to the number density of dead cells before they are cleared.

A key feature of the WAF model takes into account the observation that a normal EC in contact with other ECs on all sides does not divide, and only non-senescent cells that border holes can proliferate; the expression $\lambda_i m_i h$ ($0 \leq i \leq N - 1$) for the rate of cell division satisfies this requirement. The proliferation parameter λ_i depends on cell generation i and is expected to decrease with i (that is, it becomes harder for cells to proliferate as they get older). Also, every viable cell has some probability of dying at any time, and the death rate is assumed to be proportional to the cell number density, that is, $k_i m_i$ ($0 \leq i \leq N$). As with λ_i, the death parameter k_i depends on i but is expected to increase with i. The assumed dependences of proliferation and death-rate parameters, λ_i and k_i, are shown in Fig. 10.3(a). Cellular senescence is therefore represented in the model via three mechanisms: the Hayflick limit, the dependence of λ_i and of k_i on cell generation i. The model considers the case where viable ECs and EPCs that have homed on the wall do not undergo migration (by diffusion or transport) on the endothelium surface.

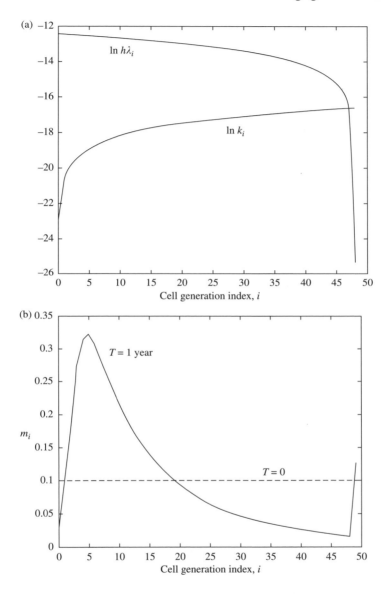

Fig. 10.3 (a) Variations of cell replication-rate coefficients ($h\lambda_i$) and death-rate coefficients (k_i) with cell generation. The replication-rate parameters λ_i range from zero ($\lambda_N = 0$ for senescent cells) to a maximum of λ_0 ($= 4\times10^{-5}$ s^{-1} for progenitor cells), and with intermediate values determined from the equation $\lambda_i = (\lambda_0 + 1)^{(N-i)/N}-1$. The cell death-rate coefficients range from a minimum of zero (for k_0) to a maximum value of 6×10^{-8} s^{-1} (for k_N), with intervening values given by the equation $k_i = (k_N + 1)^{i/N}-1$. (b) Figure demonstrating that due to the variations in (a), a sharply peaked distribution (*black solid curve*) is spontaneously generated, even from an initially flat distribution (*dashed line*). The spike at the end of the *solid curve* is due to the accumulation of senescent and dead cells.

The 53 dynamical equations of the model are given by:

$$\frac{\partial m_0}{\partial t} = \gamma h - \lambda_0 m_0 h - k_0 m_0, \tag{10.1}$$

$$\frac{\partial m_i}{\partial t} = 2\lambda_{i-1} m_{i-1} h - \lambda_i m_i h - k_i m_i \quad (1 \le i \le N-1), \tag{10.2}$$

$$\frac{\partial m_N}{\partial t} = 2\lambda_{N-1} m_{N-1} h - k_N m_N, \tag{10.3}$$

$$\frac{\partial m_{N+1}}{\partial t} = D\nabla^2 m_{N+1} + \sum_{i=0}^{N} k_i m_i - \delta m_{N+1}, \tag{10.4}$$

$$\frac{\partial h}{\partial t} = \delta m_{N+1} - \gamma h - \sum_{i=0}^{N-1} \lambda_i m_i h. \tag{10.5}$$

The first term on the right-hand side of eqn 10.1 is the EPC homing rate with a rate coefficient γ. The factor 2 in eqns 10.2 and 10.3 is due to cell doubling after division. Note that there is no cell-division term in eqn 10.3 since senescent cells do not proliferate. The first term on the right-hand side of eqn 10.4 takes into account that dead cells – prior to being cleared – become loosely attached to the endothelial wall and can diffuse laterally (D is the diffusion coefficient); the second term accounts for the death of all viable cells, and the last term is the rate of clearing dead cells, with a rate coefficient of δ.

The simulations presented in the paper of Wang *et al.* (2007) involve a square R of 1 cm^2 area on the endothelial wall

$$R = \{(x,y)|0 \le x \le 1, \ 0 \le y \le 1\},$$

and the following set of periodic boundary conditions on the number density of dead cells:

$$\begin{array}{ll} m_{N+1}(0,y,t) = m_{N+1}(1,y,t), & \nabla m_{N+1}(0,y,t) = \nabla m_{N+1}(1,y,t) \\ m_{N+1}(x,0,t) = m_{N+1}(x,1,t), & \nabla m_{N+1}(x,0,t) = \nabla m_{N+1}(x,1,t). \end{array} \tag{10.6}$$

On a macroscopic scale it is assumed that the ECs and holes are initially distributed so that the total number density is the same at every point, that is,

$$h(x,y,0) + \sum_{i=0}^{N+1} m_i(x,y,0) = \text{constant}, \ A. \tag{10.7}$$

Since one EC covers an area of the order of 100 μm^2, A is of the order of 10^6 cells per cm^2 area. One can show that

$$\iint_R \left(h + \sum_{i=0}^{N+1} m_i \right) dx dy = \text{constant}, \ A \quad \text{for all } t > 0. \tag{10.8}$$

By dividing both sides of eqn 10.7 by A, the number densities are transformed into dimensionless variables; from hereon, h and m_i ($i = 1, \ldots, N + 1$) refer to the dimensionless number densities and the A in eqn 10.7 is set to 1. Note that if $h(x,y,0)$ and $m_i(x, y, 0)(i = 1, \ldots, N+1)$ are all constants, then the system eqns 10.1–10.5 becomes a system of ordinary differential equations.

The distribution $G(t) = \{m_i(t)\}$ is referred to as the 'cell-generation profile' of the endothelium at time t. It is assumed that, on average, the endothelium of an individual at any age is composed of a distribution G of finite spread. Interestingly, due to the dependence of the cell-division and death rate-constants (λ_i and k_i) on cell generation i (see Fig. 10.3(a)), the model spontaneously generates $G(t)$ profiles that look like Gaussian distributions (see Fig. 10.4); that is, even an initially flat distribution eventually becomes Gaussian-like given enough time (see Fig. 10.3(b)). In the WAF model simulations, the chronological age of an individual is measured by the independent variable t in the dynamical eqns 10.1–10.5. If the peak of the distribution G is taken as a measure of the endothelium's age, the model simulations demonstrate that the endothelium does not have to age at the same rate as the individual's chronological age (see Fig. 10.4). Currently, there are no published measurements that permit one to correlate an individual's chronological age to the cell generation index i^* where G peaks (the simulations shown in Fig. 10.4 arbitrarily assumes that a 20-year old has a G that peaks at $i^* = 6$ (see leftmost curves in Fig. 10.4 (a)–(c)).

As shown in Fig. 10.4, the speed by which G moves towards senescence depends sensitively on the value of the EPC homing parameter γ. This parameter was varied between 0 and 1, and it was found that G(t) becomes slower if γ is increased, and faster if γ is decreased. For $\gamma = 10^{-9}$, the results are very similar to those of Fig. 10.4(a). For large values of $\gamma(>10^{-7})$, some interesting results are observed. When $\gamma = 10^{-5}$, a stationary G is reached with $i^* = 27$ at the age of $t = 60$ (see Fig. 10.4(b)). For $\gamma = 10^{-1}$ (see Fig. 10.4(c)), G moves forward (increasing i^*) for some time and then *back* to a stationary G with $i^* = 9$; clearly, this high value of γ is unrealistic. A biologically sensible value of the homing parameter γ ranges from 10^{-9} to 10^{-8} s^{-1}.

The model was also used to investigate the healing time of a wound of the endothelium of young, middle-aged, and old persons. As exhibited in Fig. 10.5, the healing time dramatically increases towards old age (compare 40 and 60 year-olds in Fig. 10.5(a)), and sensitively depends on the value of the EPC homing-rate parameter γ.

10.3 Asymmetric stem-cell division

Molecular markers for stem cells have been identified, and specific locations of stem cells in some tissues have been located. For example, hematopoietic stem cells are known to reside in the bone marrow, and neural stem cells have recently been found in certain areas of the brain (the discovery of these neural stem cells was a surprise because it was thought for a long time that neurons do not regenerate).

Pools of stem cells contain small numbers of these cells. Stem cells are thought to divide only when the need to regenerate arises. When stem cells divide, it is expected that new stem cells must be born in addition to the progeny cells that go on to

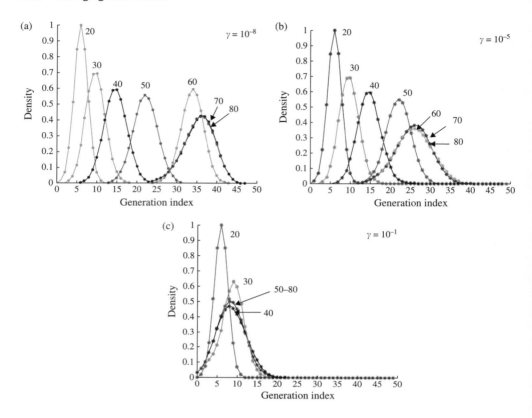

Fig. 10.4 Cell-generation profiles for a person with advancing chronological age of $T = 20$, 30, 40, 50, 60, 70, and 80 years (numbers shown near the peaks of the curves). The three values used in the simulations for the progenitor cell homing parameter γ are indicated; values of all other parameters are: $D = 10^{-8}$ cm^2 s^{-1}, $k_0 = 0$ s^{-1}, $k_N = 0.6 \times 10^{-7}$ s^{-1}, $\lambda_0 = 0.4 \times 10^{-4}$ s^{-1}, $\lambda_N = 0$ s^{-1}, $\delta = 6 \times 10^{-7}$ s^{-1}. The points on the right edge of the plots (at $i = 50, 51$) are the densities of senescent and dead cells.

differentiate. Thus, the common thinking is that a stem cell divides asymmetrically – that is, one daughter cell is identical to its mother, while the other is different from the mother and goes on to differentiate. If all stem cells give rise to twins that differentiate, then the stem-cell pool will ultimately get depleted. The asymmetry with regards to production of both stem-cell and differentiating progenies can arise by the two possibilities depicted in Fig. 10.6.

The picture on the left assumes that the progenies are twins that resemble their stem cell mother, but it is the asymmetry of the environments where the twins reside that determine the asymmetry of their fates (one niche giving rise to a stem cell, while the other gives rise to a differentiated cell). The picture on the right hypothesizes

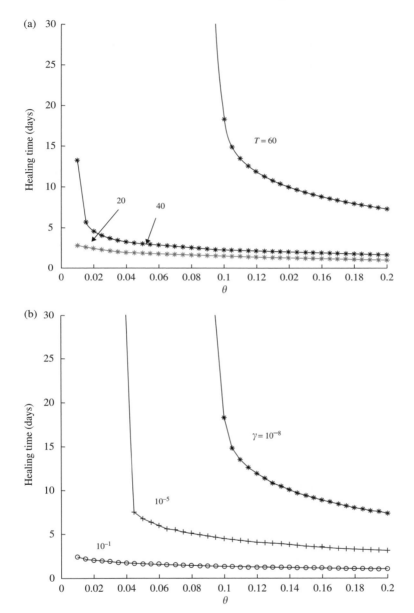

Fig. 10.5 (a) Starting with identical circular hole-density profiles, $h(x, y, 0) = \exp[-50((x - 0.5)^2 + (y - 0.5)^2)]$, on their endothelia, the time it takes (healing time) for the hole to be reduced to a fraction θ is shown for a young ($T = 20$ years), a middle-aged ($T = 40$), and an old ($T = 60$) person. Parameter values as in Fig. 10.4, and $\gamma = 10^{-8}$. (b) Starting with the cell-generation profile of an old person ($T = 60$ years) with a damaged endothelium (hole-density profile as in (a)), the healing times are determined for the three γ values indicated. Other parameter values as in Fig. 10.4.

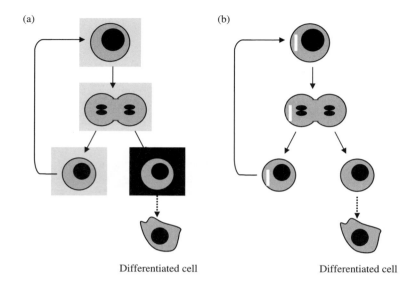

(a)   (b)

Differentiated cell   Differentiated cell

Fig. 10.6 Symmetric (a) and asymmetric (b) division of a stem-cell. In case a, the stem cell niche (*light grey background*) maintains the stemness of a daughter cell; other environments (*dark background*) direct the differentiation of a daughter cell. In case b, one daughter cell is a stem cell (identified by a *white vertical bar*) while the other goes on to differentiate. Figure adapted from Alberts *et al.* (2002).

that the asymmetry already exists internally in the daughter cells, regardless of the environment.

10.4   Maintaining the stem-cell reservoir

10.4.1   The Roeder–Loeffler model

Roeder and Loeffler (2002) proposed a dynamic model of blood stem-cell organization using the concept of '*within-tissue plasticity.*' The model assumes that there are two space compartments where hematopoietic stem cells can be found. These two compartments are depicted in Fig. 10.7. As can be seen in this figure, the hematopoietic stem cells (HSCs) are either found attached or unattached to stroma cells. Attached HSCs have been observed to be quiescent (arrested in G1 of the cell cycle) while unattached HSCs can proliferate, and *may* eventually differentiate and exit the bone marrow into the blood stream or lymphatic system. The Roeder–Loeffler model assumes that unattached HSCs may attach themselves again with stroma cells and become quiescent; this is the basis of the 'within-tissue plasticity' modelled of Roeder and Loeffler. The Roeder–Loeffler model is based on this picture of the bone marrow. The model is depicted in Fig. 10.8.

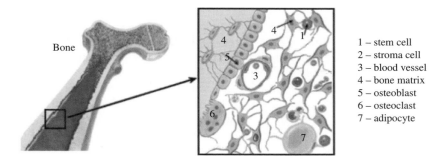

Fig. 10.7 A schematic diagram of the components of the stem-cell microenvironment in the bone marrow. Figure reproduced with permission from copyright owner cited above.

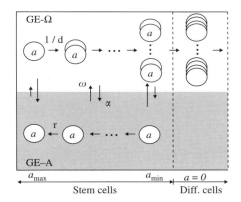

Fig. 10.8 The Roeder–Loeffler stem-cell model. Cell environments GE-A and GE-Ω represent stroma-attached or stroma-unattached situations, respectively. Cells can proliferate in GE-Ω as depicted by the overlapping circles. Attachment affinity a decreases by a factor of $1/d$ per time step in GE-Ω, and increases by a factor r per time step in GE-A. Shuttling of cells between GE-A and GE-Ω occur with rate coefficients α and ω, both quantities being dependent on a. When a goes below a threshold value $a_{\min}$, the cell exits GE-Ω and is considered differentiated. Figure reproduced with permission from Roeder and Loeffler (2002).

HSCs can be found in either growth environment GE-A or GE-Ω, corresponding to whether the HSCs are attached or unattached to stroma cells, respectively. A stem cell is characterized by two properties: the cycling status c, and the attachment affinity a. Cells in GE-A do not proliferate while those in GE-Ω do, with a turnover time of τ_c. The cycling status c has the range $0 \le c \le \tau_c$. Increasing a means increasing affinity of the cell to GE-A; it also means an increasing probability (as measured by the transition coefficient α) for a cell in GE-Ω to move to GE-A, or a decreasing transition probability (as measured by the coefficient ω) for a cell in GE-A to move to GE-Ω. Cells in GE-A are assumed to have increasing values of a with time. Cells in GE-Ω have decreasing values of a with time. Cells in GE-Ω whose a values go below a threshold value, $a_{\min}$, are considered differentiated and are irreversibly detached from GE-A so that they eventually exit the stem-cell niche (GE-Ω + GE-A).

The transition probabilities between the two growth environments are both functions of a and number of stem cells in each environment (N^A and N^Ω in GE-A and GE-Ω, respectively), i.e. $\alpha = \alpha(a, N^A)$ and $\omega = \omega(a, N^\Omega)$. These functions are shown in Fig. 10.9 where, also, a summary of the Monte-Carlo-type simulation procedure performed by Roeder and Loeffler is summarized.

The important results of the Monte-Carlo simulations using the Roeder–Loeffler model are shown in Fig. 10.9(c). The differentiation parameter d (where $1/d$ is the fraction that the affinity a decreases to after each time unit, $d > 1$, for cells in GE-Ω) and the regeneration parameter r (where r is the multiplicative factor that the affinity a increases by for each time unit, $r > 1$, for cells in GE-A) are shown to be

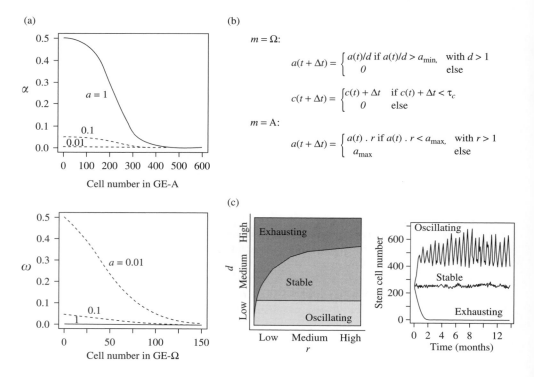

Fig. 10.9 (a) Variation of the transition intensities α and ω as functions of cell numbers and attachment affinity a. (b) Discrete dynamics of the attachment affinities in GE-A and GE-Ω, and the progression of the position of cells in the cell cycle, c. (c) Phase diagram showing the qualitative behavior of the system (stem-cell number versus time) as a function of the model parameters d and r. Figures reproduced with permission from Roeder and Loeffler (2002).

determinants of the temporal behavior of stem-cell numbers ($N^A + N^\Omega$). The figure shows that medium values of d and high values of r lead to the maintenance of (stable) stem-cell numbers. Also, as one would expect, low regeneration (low r) and high differentiation rates (high d) lead to exhaustion of the stem-cell pool. Interestingly, oscillations in stem-cell numbers characterized by large amplitudes (Fig. 10.9(c)) arise when d is small. Roeder and Loeffler speculate that these oscillations can explain cyclic neutropenia or other stem-cell disorders.

10.4.2 A deterministic model

Here, the deterministic PDE model formulated by Roeder (2003) will be described. The model allows the description of the temporal changes of densities of cells with varying attachment affinities in both GE-A and GE-Ω. The cell density at affinity a and time t, $n(a, t)$, is defined so that the number of cells in the affinity interval $[a_1, a_2]$ at time t is

$$N([a_1, a_2], t) = \int_{a_1}^{a_2} n(a, t)\mathrm{d}a. \tag{10.9}$$

The total stem-cell number is therefore

$$N(t) = N([a_{\min}, a_{\max}], t) = \int_{a_{\min}}^{a_{\max}} n(a, t)\mathrm{d}a. \tag{10.10}$$

With superscripts A and Ω referring to quantities associated with GE-A and GE-Ω, the complete system of PDE equations in $\{a_{\min} < a < a_{\max}, \ t > 0\}$ with and initial and boundary conditions is

$$\frac{\partial}{\partial t} n^A + \frac{\partial}{\partial a}(n^A \cdot v^A) = -\omega n^A + \frac{1}{\kappa} a n^\Omega, \tag{10.11}$$

$$\frac{\partial}{\partial t} n^\Omega + \frac{\partial}{\partial a}(n^\Omega \cdot v^\Omega) = -\frac{1}{\kappa} a n^\Omega + \omega n^A + 2\frac{\ln 2}{\tau_c} n^\Omega, \tag{10.12}$$

$$n^A(a, 0) = g^A(a), \tag{10.13}$$

$$n^\Omega(a, 0) = g^\Omega(a), \tag{10.14}$$

$$n^A(a_{\min}, t) = 0 , \quad n^\Omega(a_{\max}, t) = 0, \tag{10.15}$$

where, by Fig. 10.9(b) (and $v^A = (\mathrm{d}a/\mathrm{d}t)^A$, $v^\Omega = (\mathrm{d}a/\mathrm{d}t)^\Omega$)

$$v^A = a \ln r , \quad r > 1, \tag{10.16}$$

$$v^\Omega = -a \ln d , \quad d > 1, \tag{10.17}$$

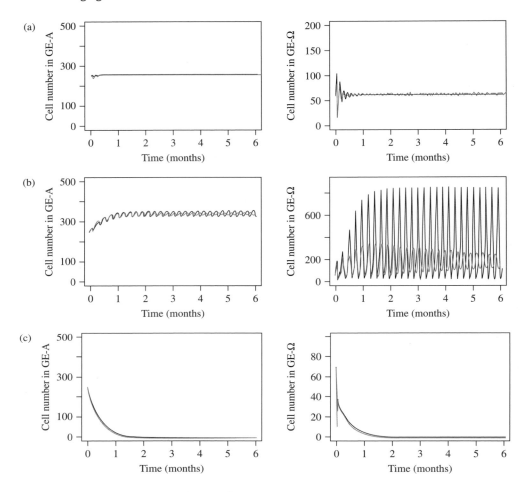

Fig. 10.10 (a) Dynamics of N^A and N^Ω simulated using the Monte-Carlo method (*red curve*, averaged over 100 simulations) and the deterministic PDE model (*black curve*) of Roeder–Loeffler. $d = 1.07$, $1/\kappa = 0.42$. Initial conditions: $N^A(0) = 250$ cells, $N^\Omega(0) = 60$ cells; $g^A(a)$ is uniform over a [0.9, 1] and $g^\Omega(a)$ is uniform over [0.01, 0.1]. (b) $d = 1.04$, $1/\kappa = 0.42$. Initial conditions as in (a). (c) $d = 1.2$, $1/\kappa = 0.1$. Initial conditions as in (a). Figures reproduced with permission from Roeder (2003). Courtesy of Ingo Roeder. (See Plate 7)

$$\omega = \frac{a_{\min}}{a} f_\omega(N^\Omega), \tag{10.18}$$

$$\alpha = \frac{a}{a_{\max}} f_\alpha(N^A). \tag{10.19}$$

The factor $(1/\kappa)$ in eqns 10.11 and 10.12 corresponds to the fraction of the G1 phase in the total cell-cycle duration. The rightmost term in eqn 10.13 is due to cell division with the rate constant equal to $(\ln 2/\tau_c)$ where τ_c is the population doubling time. The initial distributions of cells with respect to the attachment affinity a in GE-A and GE-Ω are given by the $g^A(a)$ and $g^\Omega(a)$, respectively, as shown in eqns 10.13 and 10.14. The parameter r in eqn 10.16 is referred to as the 'regeneration parameter' and the parameter d in eqn 10.17 as the 'differentiation parameter.' The functions f_ω and f_α in eqns 10.18 and 10.19 – corresponding to the detachment and attachment of stem cells from the stroma, respectively – are assumed to have the following general sigmoid form

$$f(N^e) = \frac{1}{\nu_1 + \nu_2 \exp\left(\nu_3 \frac{N^e}{\widetilde{N}}\right)} + \nu_4 \ , \quad e = \text{A or } \Omega, \tag{10.20}$$

where N^e is the number of stem cells in the growth environment e (A or Ω), $\widetilde{N}$ is a scaling factor for N^e (e.g. $\widetilde{N} = N^A + N^\Omega$) and the ν_is are adjustable parameters. See Fig. 10.9 for sample plots of this sigmoid function used in the Monte-Carlo simulations.

Roeder (2003) showed that the PDE model discussed above can be made to agree with the results of the MC method presented earlier. Figure 10.10 shows a comparison of these methods when parameters are used that give stable, oscillating and 'exhausting' stem-cell numbers.

References

Alberts, B., Johnson, A., Lewis, J., Raff, M., Roberts, K. and Walter, P. (2002) *Molecular biology of the cell*, 4th edn. Garland Science, New York.

Roeder, I. (2003) 'Dynamical Modeling of Hematopoietic Stem Cell Organization', PhD Thesis, University of Leipzig, Germany.

Op den Buijs, J., Musters, M., Verrips, T., Post, J. A., Braam, B. and van Riel, N. (2004). 'Mathematical modeling of vascular endothelial layer maintenance: the role of endothelial cell division, progenitor cell homing, and telomere shortening', *American Journal of Physiology, Heart and Circulatory Physiology* **287**, H2651–H2658.

Roeder, I. and Loeffler, M. (2002) 'A novel dynamic model of hematopoietic stem cell organization based on the concept of within-tissue plasticity,' *Experimental Hematology* **30**, 853–861.

Wang, Y., Aguda, B. D. and Friedman, A. (2007) A continuum mathematical model of endothelial layer maintenance and senescence. *Theoretical Biology and Medical Modelling* **4**, 30.

Exercises

1. Prove eqn 10.8.
2. Consider the model eqns 10.1–10.5 and the case of spatially uniform distribution of each m_i and of h. Does a steady-state generation profile always exist? If so, plot this profile (G vs. i) and determine whether it is stable.

3. Consider the system eqns 10.11–10.19 with eqn 10.20. The characteristic curves of eqns 10.11 and 10.12 are given by $da/dt = \nu^A$ and $da/dt = \nu^\Omega$. Prove that $n^A(a,t) \geq 0$, $n^\Omega(a,t) \geq 0$ for all (a,t).

4. For the system eqns 10.11–10.19 with eqn 10.20 prove that if $N^A + N^\Omega \leq$ constant for all $t > 0$, then $\int\limits_0^\infty N^\Omega(t)dt < \infty$; this suggests that $N^\Omega(t) \to 0$ as $t \to \infty$. Explain the biological implication of this.

11
Multiscale modelling of cancer

The development of cancer involves various temporal and spatial scales, as well as multiple levels of molecular and biological organization – from genes to cells, to tissue, and to the organism. The spatial scales range from subcellular to the cellular, and macroscopic or tissue level; temporal scales can range from milliseconds (e.g. activation of enzymes, or cellular signal transduction) to months or years of tumor growth. The hallmarks of cancer are briefly summarized in the first section of this chapter, followed by a discussion of modelling tumor spheroid growth (the typical model system for avascular tumors) and a model of colorectal cancer (a vascularized tumor). These two cancer models are hybrids of discrete and continuous dynamical models illustrating how the subcellular, cellular, cell population, and tissue dynamics are interfaced. The chapter ends with a discussion of continuum models that focus on the growth of the tumor boundary and treat the cell population as a continuous fluid.

11.1   Attributes of cancer

Cancer originates from cells in tissues that have acquired abnormal rates of proliferation that are not constrained by the developmental or tissue-maintenance plans of an organism. There are more than a hundred known types of human cancers, broadly categorized according to tissue of origin. *Carcinomas* begin with epithelial cells; *sarcomas* arise from connective tissues, muscles and vasculature; *leukemias* and *lymphomas* are cancers of the hematopoietic (blood) and immune systems, respectively; *gliomas* are cancers of the central nervous system, including the brain; and *retinoblastomas* are those of the eyes.

The proliferative advantage of cancer cells over normal cells can be due to any or a combination of the following attributes of cancer cells (Hanahan and Weinberg, 2000):

1. Self-sufficiency in growth signals due to the ability of cancer cells to synthesize factors that they secrete and stimulate their own division (autocrine signalling).
2. Insensitivity to antigrowth signals. During an organism's normal development, certain cells are instructed to exit the cell cycle and differentiate; in contrast, many cancer cells are not responsive to these cell-cycle-arrest signals for various reasons, including mutations of genes whose protein products are members of cell-cycle-arrest pathways.
3. Evasion of apoptosis. Cancer cells avoid dying by apoptosis due to faulty transduction of death signals.

4. Limitless replicative potential. As was discussed in Chapter 10, cellular senescence is another built-in program that safeguards the organism from cancer. However, cancer cells may override senescence by, for example, overexpressing the telomerase enzyme that maintain telomere lengths.

5. Sustained angiogenesis. The growth of a tumor involves the co-operation of other non-cancer cells. In angiogenesis, for example, endothelial cells are recruited towards the tumor to form the lining of blood vessels supplying oxygen and nutrients required for tumor growth.

6. Ability to invade tissues and metastasize. Metastasis, or the spread of tumor cells to other parts of the body, is facilitated by the tumor cells' ability to free themselves from cell–cell and cell–substrate adhesions.

It is not necessary that all of the above attributes be possessed by a cancer cell or that these attributes occur sequentially as listed above. It is believed that these physiological advantages are acquired progressively in a Darwinian-type of evolution of the cancer-cell population via some succession of genetic and epigenetic changes.

From the cancer hallmarks mentioned above, it is to be expected that cancer modelling involves a range of spatial and time scales. In terms of time scales, most mathematical models of solid tumors have focused on the three phases of growth, namely, avascular, vascular, and metastatic. The avascular growth phase occurs before angiogenesis; the start of the avascular phase would include the acquisition of genetic or epigenetic abnormalities at the single-cell level. A cell or a group of abnormal cells will then slowly (perhaps in the time scale of years) acquire proliferative advantage over the surrounding normal cells (for example, by enhanced autocrine signalling due to abnormal molecular pathways resulting from gene mutations). This is the 'Darwinian evolution' referred to above. In terms of spatial scales, existing mathematical models of avascular tumor growth are of two types: one is a discrete cell-population model that considers single-cell processes and rules on cell–cell interactions (using some Monte-Carlo-type computations and other cellular automata types); the other type of model is a continuum model that assumes space averaging of cell characteristics. An example of an avascular tumor-growth model that uses information from tumor spheroid experiments is discussed in the next section.

Vascular tumor-growth models include the complex process of angiogenesis, which will not be discussed in this chapter. In Section 11.3, a multiscale model of colorectal cancer is discussed in detail – this model assumes that there are already blood vessels from which oxygen diffuses. Both the vascular and avascular tumor models discussed in this chapter illustrate ways of linking genetic and protein cellular pathways (at the single-cell level) to the cell-population level. Models of metastasis are in their infancy and are not discussed in this book.

11.2 A multiscale model of avascular tumor growth

Multicellular tumor spheroids are observed during the early stages of many solid tumors. The availability of *in vitro* experimental models makes the study of tumor spheroids very popular. A cross-section of a tumor spheroid is shown in Fig. 11.1(a). *In vivo*, it is believed that one spheroid can be formed from a single cancer cell. With

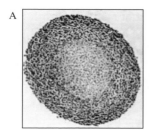
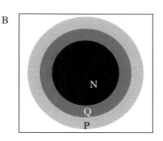

Fig. 11.1 (A) Cross-section through the center of a spheroid of tumor cells (mouse mammary tumor cells, grown from a cell line called EMT6/Ro). The spheroid diameter is ∼1200 μm. The rim of viable cells (proliferating and quiescent cells) is the *dark band* surrounding a core of necrotic cells. Picture is reproduced with permission from Jiang *et al.* (2005). Copyright 2007 Biophysical Society. (B) Schematic depiction of the necrotic (N), quiescent (Q), and proliferating (P) cell layers. The boundary between N and the viable rim is observed to be sharp, but note that the boundary between P and Q is not – there is a continuous decrease in the fraction of proliferating cells in the viable rim from the spheroid surface to N.

sufficient nutrients, the initial radial growth of the tumor is described as quasiexponential; eventually the growth saturates (i.e. reaches a steady-state radius). As depicted in Fig. 11.1, three layers of cells are formed – an outermost layer of proliferating cells (P), a middle layer of viable but quiescent cells (Q), and a core of dead or necrotic (N) cells. These layers arise due to diffusion-limited availability of nutrients from the growth medium (periphery) to the spheroid center. As demonstrated in the experimental system of Fig. 11.1, there is a clear demarcation between the viable rim (P+Q layers) and the necrotic region. Another interesting experimental observation is the approximately constant width of the viable rim as the spheroid grows. In this section, the essential features of the discrete model of Jiang *et al.* (2005) – referred to below as the Jiang model – is discussed.

The Jiang model takes the following into consideration: cell growth and division, diffusion and consumption of nutrients, production and diffusion of cell (metabolic) wastes, protein factors that promote or inhibit cell growth, intercellular adhesion, cell geometry, and cell–environment interactions.

11.2.1   Cellular scale

At the cellular scale, the Jiang model employs a discrete-lattice Monte-Carlo procedure. Three-dimensional space is partitioned into domains of cells and external medium. (A cell occupies several lattice sites to allow cell deformation.) It is assumed that a cell is in direct contact with adjacent cells (i.e. the extracellular matrix is ignored), and cell–cell interaction is through surface adhesion and competition for space. The growth dynamics of the tumor is governed by a total energy function H

that accounts for cell–cell adhesion and growth of cell volumes:

$$H = \sum_{\text{lattice sites}} [J_{S1-S2}][1 - \delta(S1, S2)] + \sum_{\text{cells}} \gamma[v - V^T]^2, \tag{11.1}$$

where the first sum (over all lattice sites) on the right-hand side of the equation gives the total adhesive energy between cells, and the second sum (over all cells) accounts for changes in cell volumes. The factor J_{S1-S2} is the adhesive energy between cells labelled S1 and S2, δ is the Kroneker delta function, V^T is the target cell volume (explained below), v is the current cell volume, and γ is the coefficient of volume elasticity. Note that the second term provides the potential for cell growth from v to V^T. The value of the parameter J_{S1-S2} varies according to the types that cells S1 and S2 belong to – this assumption embodies the *differential adhesion hypothesis*, which states that differences in adhesivity between different cell types lead to cell sorting that minimizes surface energy. The Jiang model assumes three types of cells: proliferating, quiescent, and necrotic. For computational purposes, the external culture medium is considered as a special 'cell'. Each proliferating cell has a V^T that is twice its volume at birth; thus, the cell volume has to grow to minimize H. The target volumes of a quiescent cell and a necrotic cell are equal to their respective current volumes (in other words, these cells do not grow). The space occupied by necrotic cells cannot be invaded by viable cells, but the external medium can be invaded by proliferating cells.

The Monte-Carlo procedure for minimizing H involves randomly selecting a lattice site, changing the cell ID of this site to the value of one of its neighbors with a different ID, calculating the ΔH due to this ID change, and then calculating the probability of accepting such an ID change according to the following equation:

$$p = \begin{cases} 1 & \Delta H < 0 \\ e^{-\beta \Delta H} & \Delta H \geq 0, \end{cases} \tag{11.2}$$

where β is a constant (the equation above says that the ID change is accepted if ΔH is negative, whereas the probability of accepting the change decreases exponentially with positive values of ΔH). A Monte-Carlo step (MCS) involves the number of trial lattice updates equal to the total number of lattice sites.

According to the Jiang model, in order for a proliferating cell to divide it must satisfy two criteria: that it has grown to at least the target volume V^T, and that its age (from birth) must be at least equal to the duration of a cell cycle. After cell division, one of the halves is assigned a new cell ID. A daughter cell inherits all the properties of her parent.

11.2.2 Extracellular scale

Concentrations of molecules, u_i, found in the extracellular environment evolve according to the following continuous reaction-diffusion equation:

$$\frac{\partial u_i}{\partial t} = D_i \nabla^2 u_i + R_i(x, y, z), \tag{11.3}$$

where the subscript i ranges from oxygen, nutrients, metabolic waste, growth-promoting factors, growth-inhibiting factors, and others. The function R_i is the space-dependent net production rate of molecule i, and D_i is the molecule's diffusion coefficient. The Jiang model assumes that necrotic and quiescent cells secrete growth-inhibitory factors. It is also assumed that the concentrations of oxygen, nutrients, and growth-promoting factors in the extracellular medium are homogeneous and constant in time.

11.2.3   Subcellular scale

In general, at the subcellular scale, gene- and protein-regulatory networks that control cellular physiology are considered. The activities of these networks allow further classification of cells according to their position in the cell cycle (G1, S, G2, and M). (The subcellular scale is the area of multiscale cancer modelling that is not quite developed at this time.) The Jiang model, in its original form, assumed a simplistic set of linear pathways from S-phase-promoting and -inhibiting proteins to the E2F transcription factor and will not be discussed here (the next section will discuss another more appropriate gene network).

It is common to treat protein and gene-regulatory networks as Boolean networks with Boolean dynamics, i.e. all the proteins have only two states: on or off. The Jiang model assumes that the dynamics of the intracellular Boolean network is affected by the local concentrations of growth-promoting and -inhibiting factors only (u_g and u_i, respectively), and that the switching on or off of a protein in the network depends on the value of a 'factor level' f defined as

$$f = \frac{1}{1 + e^{-\alpha(\Delta g - \theta)}}, \quad \text{where } \Delta g = (u_g - u_i)/u_g^0, \tag{11.4}$$

where θ is a factor level threshold, u_g^0 is the concentration of growth-promoting factors in the extracellular medium, and α is a constant. A protein X in the network is turned on if the factor level f is above the threshold and one of the following situations holds: either all the proteins affecting X are activating and are *on*, or all the proteins affecting X are inhibitory and are *off*. All other situations turn the protein off. If f is below the threshold, its value is interpreted as the probability that a protein will be turned on. If E2F is *off* then the cell goes into quiescence, otherwise it progresses through the cell cycle. In the Jiang model, a cell dies if $u_o < 0.02$ mM, $u_n < 0.06$ mM, and $u_w > 8$ mM (where u_o, u_n, and u_w are the concentrations of oxygen, nutrients, and wastes, respectively). The Monte-Carlo simulation scheme is shown in Fig. 11.2.

A simulation run of the Jiang model is shown in Fig. 11.3. The top panel (Fig. 11.3A) shows the cross-sections of the growing tumor from a single cancer cell to a spheroid; these simulations reproduce the well-defined layers of proliferating, quiescent, and necrotic cells observed in experiments. Figure 11.3B demonstrates the good fit between the model and experimental data on growth in tumor volume. Figure 11.3C compares model predictions and experimental data on the widths of the viable rim (proliferating and quiescent cells) and diameters of the necrotic core. The experimental data show that the width of the viable rim is approximately constant and

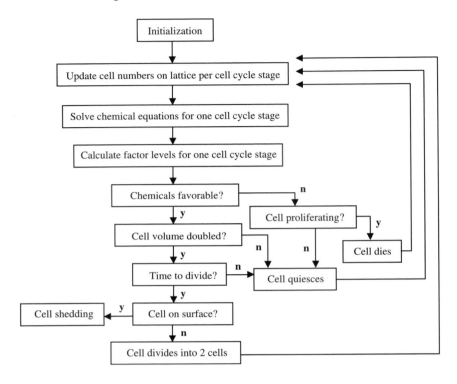

Fig. 11.2 Flowchart of the Monte-Carlo simulation steps in the Jiang *et al.* (2005) model. In this model, cell-cycle progression is represented by 16 stages (G1 and S phases have 6 stages each, and G2/M combined has 4 stages). Cell shedding has been observed experimentally for mitotic cells located on the spheroid surface. Steps in this flowchart are explained in the text. Figure is modified from Jiang *et al.* (2005).

that the necrotic core radius linearly increases for tumor spheroid diameters between 370–830 μm. The computer simulations agree well with the data at the larger diameters, but quantitatively differs in the initial stages of tumor growth.

11.3   A multiscale model of colorectal cancer

In the USA, there are over 130 000 persons diagnosed each year with colorectal cancer with about 50 000 dying of the disease. The cancer usually starts as a benign tumor (called polyps) that, under conditions allowing further genetic mutations, can progress into a full-blown malignant tumor.

Approximately 80% of colorectal cancers are sporadic (about 20% are associated with family history of colon cancers). Individuals with the hereditary colon cancer syndrome, called 'familial adenomatous polyposis', often develop hundreds or even thousands of colon polyps in their teenage years, and are therefore quite susceptible

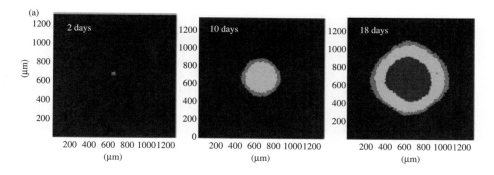

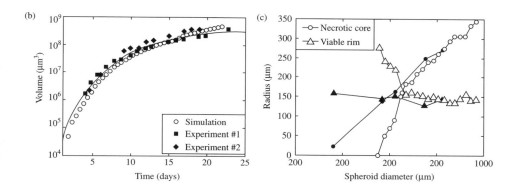

Fig. 11.3 Simulations of tumor spheroid growth using the Jiang model. (A) Cross-section of the spheroid. *Cyan, yellow, and magenta areas* represent proliferating, quiescent, and necrotic cells, respectively. The simulation was started with a single cell. (B) Growth of the spheroid volume with time. *Solid black symbols* are from experimental data. *Empty white circles* are from model simulations. The *continuous curve* is the best fit to the experimental data. (C) Model predictions (*empty circles and triangles*) and experimental data (*solid circles and triangles*) on the widths of the viable rim (proliferating and quiescent cells) and diameters of the necrotic core. All pictures are reproduced with permission from Jiang *et al.* (2005). Copyright 2007 Biophysical Society. (See Plate 8)

to developing colon cancers. In the following subsections, the different levels at which the multiscale model of Ribba *et al.* (2006) is based will be discussed in detail.

11.3.1   Gene level: a Boolean network

The five genes most commonly observed to be mutated in colorectal cancers are: APC, K-Ras, TGFβ, SMAD4, and p53. Ribba *et al.* (2006) assumed the gene network shown in Fig. 11.4 that affects cell proliferation and cell death. As indicated in this figure, these genes are involved in sensing growth factors, hypoxia (low oxygen), cell crowding

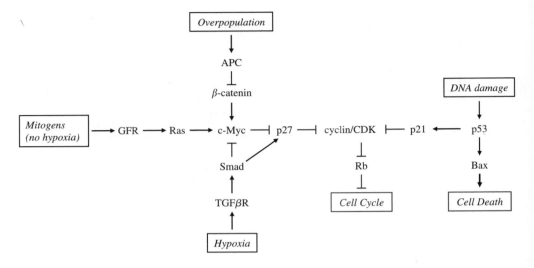

Fig. 11.4 A caricature of the gene- or protein-interaction network involved in the progression of colorectal cancer (figure adapted from Ribba *et al.* (2006) with minor modifications). This network accounts for the following observations: mitogens or growth factors promote cell proliferation in the presence of sufficient oxygen; cell crowding (overpopulation) inhibits cell proliferation; lack of oxygen (hypoxia) inhibits cell proliferation; DNA damage induces cell death as well as inhibits cell proliferation. An *arrow* means 'induces activation of', and a *hammerhead* means 'induces inhibition of'. Abbreviations: GFR = growth factor receptor, APC = adenomatous polyposis coli, TGFβR = TGFβ receptor, CDK = cyclin-dependent kinase, Rb = retinoblastoma protein.

(overpopulation), and DNA damage. This network is highly simplified and should only be viewed for the purpose of illustrating the multiscale model as simply as possible. More details about these genes are given in the caption of Fig. 11.4.

The gene network is used in the model of Ribba *et al.* (2006) as a Boolean network to make decisions whether cells continue into S-phase (proliferative mode), exit the cell cycle to a quiescent mode (G0), or die (apoptose). (The model is simply referred to as the Ribba model from hereon.) The Boolean rules corresponding to the gene network are given in Table 11.1. This table gives the logical value (0 or 1) of a gene at time $t+1$ depending on the values at time t of the genes that affect it. For example, the value of c-MYC at time $t + 1$ is equal to $(\mathbf{RAS}^t)$ **OR** $(\beta\text{-}\mathbf{CATENIN}^t)$ **OR** $(\mathbf{NOT\ SMAD}^t)$ where superscript t means the value at time t. As an example of how a signal propagates through this Boolean gene network, consider the case when the p27 gene is deleted and the dynamics of the pathway from 'DNA-damage signal' to 'cell proliferation' is followed. Assume that at time t, $\mathrm{RB}^t = 0$, $\mathrm{CYC\text{-}CDK}^t = 1$, $\mathrm{p21}^t = 0$, and $\mathrm{p53}^t = 0$; this means that cells are proliferating at time t. When an above-threshold DNA-damage signal is introduced at time t one obtains the following from the gene network: $\mathrm{p53}^{t+1} = 1$, $\mathrm{p21}^{t+2} = 1$, $\mathrm{CYC\text{-}CDK}^{t+3} = 0$, $\mathrm{RB}^{t+4} = 1$, and (cell proliferation)$^{t+5} = 0$;

Table 11.1 Logic rules corresponding to the Boolean gene network of Fig. 11.4. (Table is reproduced with permission from Ribba *et al.* (2006)).

Node	Boolean updating function
APC^t	$APC^{t+1} = \begin{cases} 1 & \textit{if Overpopulation signal} \\ 0 & \textit{Otherwise} \end{cases}$
	$APC^{t+1} = 0 \; \textit{if mutated}$
βcat^t	$\beta cat^{t+1} = \neg APC^t$
$cmyc^t$	$cmyc^{t+1} = RAS^t \wedge \beta cat^t \wedge \neg SMAD^t$
$p27^t$	$p27^{t+1} = SMAD^t \vee \neg cmyc^t$
$p21^t$	$p21^{t+1} = p53^t$
Bax^t	$Bax^{t+1} = p53^t$
$SMAD^t$	$SMAD^{t+1} = \begin{cases} 1 & \textit{if Hypoxia signal} \\ 0 & \textit{otherwise} \end{cases}$
	$SMAD^{t+1} = 0 \; \textit{if mutated}$
RAS^t	$RAS^{t+1} = \begin{cases} 1 & \textit{if no Hypoxia signal} \\ 0 & \textit{otherwise} \end{cases}$
	$RAS^{t+1} = 1 \; \textit{if mutated}$
$p53^t$	$p53^{t+1} = \begin{cases} 1 & \textit{if DNA damage signal} \\ 0 & \textit{otherwise} \end{cases}$
	$p53^{t+1} = 0 \; \textit{if mutated}$
$CycCDK^t$	$CycCDK^{t+1} = \neg p21^t \wedge \neg p27^t$
Rb^t	$Rb^{t+1} = \neg CycCDK^t$

that is, the cell cycle is arrested at time $t+5$ according to Fig. 11.4. However, this cell-cycle arrest due to the DNA-damage signal does not actually happen because at time $t+3$ the cell had already gone into apoptosis according to the figure (one must remember this when the Ribba model is being discussed below). Note that when APC, Smad, and Ras are mutated, their initial Boolean values are set to 0, 0, and 1, respectively.

11.3.2   Cell level: a discrete cell-cycle model

In the Ribba model, a cell is either in the cell cycle, in a quiescent (G0) state, or in apoptosis. In the model, the cell-cycle phases are G1, S, and G2/M (G2 and M are lumped together). Let φ denote a cell state where $\varphi \in \{$G1, S, G2/M, G0, apoptosis$\}$. These cell states and their interconversions – according to the Ribba model – are shown in Fig. 11.5.

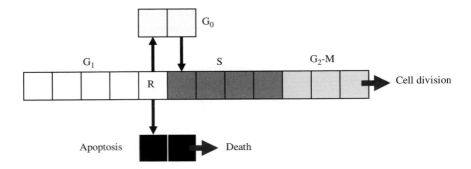

Fig. 11.5 The discrete model of cell-cycle progression used in the model of Ribba *et al.* (2006). 'R' represents the checkpoint called the 'restriction point' where a cell 'decides' to apoptose or quiesce depending on extracellular signals such as those shown in Fig. 11.4.

There is only one cell-cycle checkpoint considered in the model, namely, the restriction point (R point). At this point, the decision is made whether the cell proceeds to S phase, or to G0, or to apoptosis. This decision is governed by the genetic status of the cell – that is, whether the genes shown in the network of Fig. 11.4 are normal or not. In the Ribba model, cells that go into G0 stay there for some time and then return to the S phase of the cell cycle. Cells that have gone into apoptosis stay in this state for some time before being cleared from the system. At the end of the G2/M phase, one cell gives rise to two daughter cells in the G1 state.

The Ribba model assumes discrete cell states, each state φ having a constant time duration τ_φ; furthermore, these durations are all integer multiples of a discrete time step dt:

$$\tau_\varphi = dt \times N_\varphi \quad \text{where } \varphi \in \{G1, S, G2/M, G0, \text{apoptosis}\}. \tag{11.5}$$

Cells in the system are therefore identified by an age index a and cell state φ, where $a \in \{1, 2, \ldots N_\varphi\}$ and $\varphi \in \{G1, S, G2/M, G0, \text{apoptosis}\}$. The binary index (a, φ) will be referred to as the cell state from now on.

11.3.3   Tissue level: colonies of cells and oxygen supply

Here, a two-dimensional simulation of colorectal tumor tissue using the Ribba model is discussed. A square tissue with five circular colonies of cells and two sources of oxygen (blood vessels) make up the model domain Ω (see Fig. 11.6).

At the tissue level, the dynamical variables are the number densities (cells per unit area), denoted by $n_{a,\varphi}$, of cells with age index a and state φ. Another variable is the concentration C of oxygen that diffuses radially from the blood vessels and is then taken up by the cells according to the following equation:

$$\frac{\partial C}{\partial t} = \nabla \cdot (D\nabla C) - \sum_{a,\varphi} \alpha_\varphi n_{a,\varphi} \quad \text{on } \Omega/\Omega_{bv}, \tag{11.6}$$

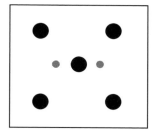

Fig. 11.6 Square domain and initial conditions of the two-dimensional tumor-growth model of Ribba *et al.* The *smaller grey discs* are blood vessels and the *black discs* are tumor-cell colonies.

with

$$C = C_{\text{max}} \quad \text{on } \Omega_{bv}$$
$$C = 0 \qquad \text{at } \partial\Omega,$$

where D is the oxygen diffusion coefficient, α_φ are the constant coefficients of oxygen uptake by cells in various states, Ω is the computational domain (Fig. 11.6), $\partial\Omega$ is this domain's boundary, and Ω_{bv} is the area covered by the blood vessels. Ω/Ω_{bv} means the computational domain excluding Ω_{bv}.

The spatiotemporal evolution of the cell number densities is described by the following system of PDEs:

$$\frac{\partial n_{a,\varphi}}{\partial t} + \nabla \cdot (n_{a,\varphi} \mathbf{v}) = P_{a,\varphi}, \tag{11.7}$$

where

$$a \in \{1, 2, \dots N_\varphi\}$$
$$\varphi \in \{\text{G1, S, G2/M, G0, apoptosis}\}.$$

The second term on the left-hand side is the advection term corresponding to the movement of cells in space, usually assumed to be in a porous medium that follows Darcy's law – a phenomenological equation relating the velocity $\mathbf{v}$ of cell movement to a local pressure gradient $\nabla\sigma$:

$$\mathbf{v} = -k\nabla\sigma, \tag{11.8}$$

where k is a permeability constant. It is assumed that the total local cell density is maintained constant throughout a growing tumor-cell colony:

$$\sum_{a,\varphi} n_{a,\varphi} = \text{constant}. \tag{11.9}$$

The term on the right-hand side of eqn 11.7, $P_{a,\varphi}$, is the net rate of change in the density of cells in the (a, φ) state. This source term is the link between the tissue level and the two other levels (i.e. cellular and genetic) in the model. For example, consider the case $\varphi^* \equiv G1$ and $a^* \equiv N_{G1}$ (the asterisks are just being used to distinguish these particular assignments of φ and a). Let $R_{(a^*-1,\varphi^*)\rightarrow(a^*,\varphi^*)}$ be the transition rate from the state $(N_{G1}-1,G1)$ to $(N_{G1},G1)$, $R_{(a^*,\varphi^*)\rightarrow(1,S)}$ be the transition rate from the state $(N_{G1},G1)$ to $(1,G0)$, and $R_{(a^*,\varphi^*)\rightarrow(1,G0)}$ be the transition rate from the state $(N_{G1},G1)$ to $(1,\text{apoptosis})$. In the Ribba *et al.* model, the expression for $P_{a*,\varphi*}$ is given by

$$P_{a^*,\varphi^*} = \begin{cases} R_{(a^*-1,\varphi^*)\rightarrow(a^*,\varphi^*)} - R_{(a^*,\varphi^*)\rightarrow(1,G0)} & \text{IF } \{(\textit{hypoxia} \text{ OR } \textit{overpopulation}) \text{ AND (NOT } \textit{die})\} \\ R_{(a^*-1,\varphi^*)\rightarrow(a^*,\varphi^*)} - R_{(a^*,\varphi^*)\rightarrow(1,\text{apoptosis})} & \text{IF } \textit{die} \\ R_{(a^*-1,\varphi^*)\rightarrow(a^*,\varphi^*)} - R_{(a^*,\varphi^*)\rightarrow(1,S)} & \text{otherwise,} \end{cases}$$

$$(11.10)$$

where *hypoxia*, *overpopulation*, and *die* are logical variables whose values are defined according to specified thresholds of local oxygen concentration (C_{th}), local cell population density (n_{th}), and DNA-damage signal (D_{th}), respectively, and to the mutation status of particular genes:

$$hypoxia = \begin{cases} True & \text{IF } \{(C < C_{th}) \text{ AND } SMAD\} \\ False & \text{otherwise,} \end{cases}$$

$$(11.11)$$

$$overpopulation = \begin{cases} True & \text{IF } \{(n > n_{th}) \text{ AND } APC\} \\ False & \text{otherwise,} \end{cases}$$

$$(11.12)$$

$$die = \begin{cases} True & \text{IF } \{(D > D_{th}) \text{ AND } P53\} \\ False & \text{otherwise.} \end{cases}$$

$$(11.13)$$

In the above definitions, *SMAD, APC,* and *P53* represent the non-mutated (or wild-type) genes (these are the only genes considered in the Ribba *et al.* model because of their direct relevance to colorectal cancer; the other genes in the network of Fig. 11.4 are all assumed to be wild type).

The set of eqns 11.5–11.13 with initial conditions shown in Fig. 11.6 and boundary condition $\partial\sigma/\partial n = 0$ completes the definition of the Ribba model. Substituting eqn 11.7 into eqn 11.9, one obtains

$$-\nabla \cdot (k\nabla\sigma) = \sum_{a,\varphi} P_{a,\varphi},$$

$$(11.14)$$

which allows the determination of the pressure field.

Numerical simulations of cell population growth using the Ribba model are shown in Fig. 11.7 for normal cells and for those having mutations in APC, SMAD, and RAS.

Normal cell-population level tapers off after some time due to overpopulation and hypoxia signals. When APC is mutated, the overpopulation-signalling pathway is cut

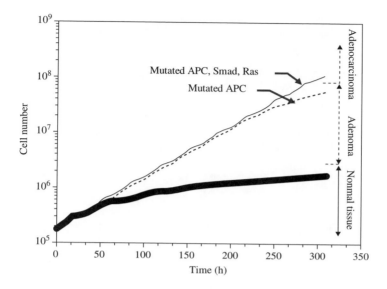

Fig. 11.7 Growth in cell numbers with time predicted by the model of Ribba *et al.* (2006). *Thick black curve* represents growth of normal cells. Growth curves for cells with gene mutations are shown and are identified with cases of adenoma and adenocarcinoma. Figure reproduced with permission from Ribba *et al.* (2006). Copyright 2006 Ribba *et al.*

off and cells are permitted to proliferate beyond normal levels (causing adenoma, as interpreted in the figure); eventually hypoxia sets in to restrain population growth. As demonstrated by the model, adenocarcinoma results when APC, SMAD, and RAS are mutated so that cell proliferation becomes unrestrained.

Figure 11.8 shows that when hypoxia occurs in the cell colonies, SMAD/RAS are activated and induce cell quiescence.

■ 11.4 Continuum models of solid tumor growth

Continuum models of solid tumor growth consider dynamical variables representing cell densities that are continuous functions of time and space, instead of discrete numbers of cells comprising the tumor. The growth models discussed in this section are examples of free-boundary problems whose general mathematical analysis is still an open mathematical problem. The instabilities associated with the morphology of the free boundary – as determined by parameters such as cell-proliferation rate and cell–cell adhesion – are thought to be relevant in the early stages of tumor invasion into the surrounding tissue.

11.4.1 Three types of cells

Following Pettet *et al.* (2001), consider a model of a growing tumor that includes three types of cells, namely, proliferating, quiescent, and dead cells; and let the mass

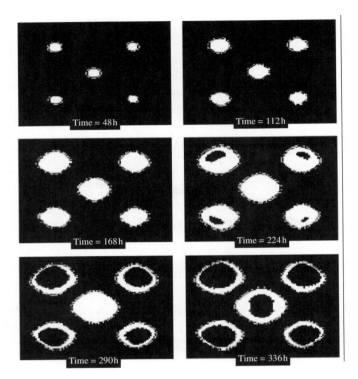

Fig. 11.8 Growth of tumor-cell colonies starting with the configuration shown in Fig. 11.6. Areas occupied by proliferating cells are shown in *white*; *black areas* surrounded by white areas are those occupied by cells that have gone into quiescence due to hypoxia and subsequent Smad and Ras activation. Figure is reproduced with permission from Ribba *et al.* (2006). Copyright 2006 Ribba *et al.*

densities of these cells be p, q, and w, respectively. Due to the birth of new cells and removal of dead cells, there is continuous cell motion with velocity $\mathbf{v}$ that depends on space and time. The equations for the conservation of cell mass densities are the following:

$$\frac{\partial p}{\partial t} + \text{div}(p\mathbf{v}) = K_{pp}(c)p + K_{pq}(c)q - K_{qp}(c)p - K_{wp}(c)p, \qquad (11.15)$$

$$\frac{\partial q}{\partial t} + \text{div}(q\mathbf{v}) = K_{qp}(c)p - K_{pq}(c)q - K_{wq}(c)q, \qquad (11.16)$$

$$\frac{\partial w}{\partial t} + \text{div}(w\mathbf{v}) = K_{wp}(c)p + K_{wq}(c)q - k_r w. \qquad (11.17)$$

The term $k_r w$ in eqn 11.17 represents the removal rate of dead cells; the other terms on the right-hand sides of eqns. 11.15–11.17 give the rates of transitions between cellular states as shown in Fig. 11.9.

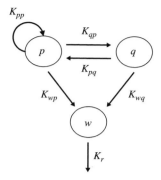

Fig. 11.9 Diagram showing the transitions between proliferating (p), quiescent (q), and dead (w) cells. The symbols adjacent to each transition arrow are the rate coefficients.

The rate coefficient $K_{ij}(c)$, for the transition from state j to i, is assumed to depend on the local concentration c of a nutrient. For *avascular* tumors, this nutrient diffuses in the tumor region $\Omega(t)$ and is consumed by proliferating and quiescent cells. The evolution of c is modelled by the following equation:

$$\varepsilon_0 \frac{\partial c}{\partial t} = D_c \nabla^2 c - \lambda(p, q)c \quad \text{in } \Omega(t). \tag{11.18}$$

The nutrient's diffusion coefficient, D_c, is assumed constant. The constant ε_0 corresponds to the ratio between the time scale of nutrient diffusion and of tumor growth, and is therefore small in magnitude. The rate coefficient, $\lambda(p, q)$, of nutrient consumption is a function of the densities of viable cells, p and q. (In the discussion below, λ is assumed constant.)

As in the Ribba *et al.* model, the growing tumor is viewed as a fluid with velocity $\mathbf{v}$ moving in a porous medium that follows Darcy's law:

$$\mathbf{v} = -\beta \nabla \sigma, \tag{11.19}$$

where σ is the fluid pressure and β is a positive constant. A further assumption is that all cells are identical in volume and mass, and that the total cell density at any location within the tumor at any time is a constant:

$$p + q + w = \text{ constant, } C. \tag{11.20}$$

For simplicity, let $\beta = 1$ and $C = 1$. From the sum of eqns 11.15–11.17, one obtains

$$\text{div}\mathbf{v} = K_{pp}(c)p - k_r w. \tag{11.21}$$

Replacing eqn 11.17 with eqn 11.21, and substituting $w = 1 - p - q$, one arrives at the following system of equations (D_c is set to 1):

$$\frac{\partial p}{\partial t} - \nabla \sigma \cdot \nabla p = f(c, p, q) \quad \text{in } \Omega(t),\ t > 0, \tag{11.22}$$

$$\frac{\partial q}{\partial t} - \nabla \sigma \cdot \nabla q = g(c, p, q) \quad \text{in } \Omega(t),\ t > 0, \tag{11.23}$$

$$\nabla^2 \sigma = -h(c, p, q) \quad \text{in } \Omega(t),\ t > 0, \tag{11.24}$$

$$\varepsilon_0 \frac{\partial c}{\partial t} = \nabla^2 c - \lambda c \quad \text{in } \Omega(t), t > 0, \tag{11.25}$$

where

$$f(c, p, q) = K_{pp}(c)p + K_{pq}(c)q - K_{qp}(c)p - K_{wp}(c)p - h(c, p, q)p$$
$$g(c, p, q) = K_{qp}(c)p - K_{pq}(c)q - K_{wq}(c)q - h(c, p, q)q$$
$$h(c, p, q) = K_{pp}(c)p - k_r(1 - p - q).$$

Let $\partial\Omega(t)$ denote the boundary of the growing tumor; here, the boundary conditions on the nutrient concentration, 'fluid' pressure, and 'fluid' velocity normal to the boundary (v_n) are taken to be the following:

$$c = \bar{c} \quad \text{on } \partial\Omega(t), t > 0, \tag{11.26}$$

$$\sigma = \gamma\kappa \quad \text{on } \partial\Omega(t), t > 0, \tag{11.27}$$

$$v_n \equiv \mathbf{v} \cdot \mathbf{n} = -\frac{\partial \sigma}{\partial n} \quad \text{on } \partial\Omega(t), t > 0, \tag{11.28}$$

where $\mathbf{n}$ is the normal vector, κ is the boundary's mean curvature, and γ is the surface-tension coefficient. For a sphere of radius R, $\kappa \equiv \frac{1}{R}$. What keeps the cells together in a solid tumor is assumed to be the surface tension attributed to cell-to-cell adhesiveness with strength proportional to γ. Finally, the following initial conditions complete the definition of the mathematical problem:

$$\Omega(0) = \Omega_0, \tag{11.29}$$

$$p(x, 0) = p_0(x) \geq 0,\ q(x, 0) = q_0(x) \geq 0,\ p_0(x) + q_0(x) \leq 1 \text{ in } \Omega_0, \tag{11.30}$$

$$c(x, 0) = c_0(x) \geq 0 \quad \text{in } \Omega_0. \tag{11.31}$$

Chen and Friedman (2003) proved the existence of a unique smooth solution to the system 11.22–11.31 for some finite time interval $0 \leq t \leq T$ as long as the initial dataset $\{p_0, q_0, c_0\}$ is sufficiently smooth and the initial and boundary conditions are consistent with the differential equation for c at $\partial\Omega(0)$. In general, one cannot extend the solution beyond some finite time T. However, for the special case of radially symmetric initial data, the existence of a solution of eqns 11.22–11.31 for all $t > 0$ has been proven by

Cui and Friedman (2003) under some conditions on the parameters, as stated in the following theorem:

Theorem 11.1.A *(Cui and Friedman, 2003)*
There exists a unique radially symmetric solution for the system (eqns 11.22–11.31) for all $t > 0$ when the following conditions hold:

$$\varepsilon_0 = 0, \tag{11.32}$$

$$K'_{pp}(c) > 0, \quad K'_{pq}(c) > 0, \quad K'_{wp}(c) \le 0, \quad K'_{wq}(c) < 0,$$
$$K'_{qp}(c) < 0, \quad K'_{pp}(c) + K'_{wq}(c) > 0 \quad (0 \le c \le \bar{c}), \tag{11.33}$$

$$K_{pp}(0) = 0, \quad K_{pq}(0) = 0, \quad K_{wp}(\bar{c}) = 0, \quad K_{wq}(\bar{c}) = 0, \quad K_{qp}(\bar{c}) = 0. \tag{11.34}$$

Furthermore, the free boundary $\partial\Omega(t) = \{r = R(t)\}$ satisfies the following inequalities:

$$\rho_1 \le R(t) \le \rho_2 \quad for \ all \ t > 0, \tag{11.35}$$

where ρ_1 and ρ_2 are positive constants.

Some remarks on the biological relevance of the conditions of the preceding theorem can be made. Equation 11.32 states the assumption that the nutrient concentration spatial profile is steady in time. The dependences of the rate coefficients K_{ij}s on c – as presented in eqn 11.33 – are intuitively clear; for example, $K'_{pp}(c) > 0$ is consistent with the fact that cell-proliferation rate increases with nutrient concentration, and $K'_{qp}(c) < 0$ with the fact that there is increasing rate to quiescence when c decreases. The condition $K'_{pp}(c) + K'_{wq}(c) > 0$ states that the rate coefficient for cell proliferation is larger in magnitude than the rate coefficient for death of quiescent cells; this assumption is based on experimental data (Dorie *et al.* 1986; Dorie *et al.* 1982). Lastly, eqn 11.35 asserts that the radially symmetric tumor does not grow beyond a finite radius of ρ_2. The calculation of ρ_2 in terms of experimentally accessible parameters would be of practical significance. It can be shown that, for radially symmetric tumor growth, $R(t)$ satisfies the following integrodifferential equation:

$$R^2(t)\frac{dR(t)}{dt} = \int_0^{R(t)} h(c, p, q)r^2 dr. \tag{11.36}$$

Solving eqn 11.36 is a difficult problem because the unknown variables $\{c, p, q\}$ cannot be decoupled from the variable σ due to eqns 11.22–11.24. Friedman (2006) discussed several open mathematical problems concerning the system 11.22–11.31, including the existence of solutions that are not radially symmetric, the stability of solutions, as well as the extension of Theorem 11.1.A to the case where $\varepsilon_0 > 0$. Some of these open problems have been solved for the special case of a single type of cells, as discussed next.

11.4.2 One type of cells

Here, only proliferating cells are assumed to represent the growth of a tumor. The right-hand side of eqn 11.15 simplifies to:

$$\frac{\partial p}{\partial t} + \operatorname{div}(p\mathbf{v}) = K_{pp}(c)p - K_{wp}(c)p. \tag{11.37}$$

Since cell density is still kept constant ($p=1$), the right-hand side of the latter equation is now just a function of c, and the left-hand side simplifies to $\operatorname{div}\mathbf{v}$, so that

$$\operatorname{div}\mathbf{v} = K_{pp}(c) - K_{wp}(c) \equiv S(c). \tag{11.38}$$

A biologically feasible form of the function $S(c)$ is:

$$S(c) = \mu(c - \tilde{c}), \tag{11.39}$$

where μ is a positive constant and $\tilde{c}$ is some threshold nutrient concentration above which cells proliferate leading to the expansion of the tumor, and below which the tumor shrinks. Application of Darcy's law, and non-dimensionalizing, give

$$\nabla^2 \sigma = -S(c). \tag{11.40}$$

The system of equations for the case of one type of cells (proliferating cells) is summarized below:

$$\frac{\partial c}{\partial t} - \nabla^2 c + c = 0 \quad \text{in } \Omega(t),\ t > 0, \tag{11.41}$$

$$\nabla^2 \sigma = -\mu(c - \tilde{c}) \quad \text{in } \Omega(t),\ t > 0, \tag{11.42}$$

$$c = \bar{c} \quad \text{on } \partial\Omega(t),\ t > 0 \text{ and } \bar{c} > \tilde{c}, \tag{11.43}$$

$$\sigma = \gamma\kappa \quad \text{on } \partial\Omega(t),\ t > 0, \tag{11.44}$$

$$\frac{\partial \sigma}{\partial n} = -v_n \quad \text{on } \partial\Omega(t),\ t > 0, \tag{11.45}$$

with initial conditions

$$\Omega(0) = \Omega_0, \tag{11.46}$$

$$c|_{t=0} = c_0(x) \quad \text{for } x \in \Omega_0. \tag{11.47}$$

Note that the equations are non-dimensionalized except those involving the growth-rate parameter μ and surface-tension coefficient γ. It turns out, as shown below, that the value of the ratio (μ/γ) plays an important role in deciding the stability of the tumor boundary.

It has been shown (Bazaliy and Friedman, 2003) that the system 11.41–11.47 has a unique solution for a small time interval. As in eqn 11.36, one can show that radially symmetric solutions satisfy the following equation

$$R^2(t)\frac{\mathrm{d}R(t)}{\mathrm{d}t} = \int_0^{R(t)} \mu(c(r,t) - \tilde{c})r^2 \mathrm{d}r. \tag{11.48}$$

The following theorem, proven by Friedman and Reitich (1999) establishes the existence of a unique radially symmetric stationary solution.

Theorem 11.1.B *(Friedman and Reitich, 1999)*
There exists a unique radially symmetric stationary solution of system 11.41–11.47 given by

$$c_s(r) = \bar{c}\frac{R_s}{\sinh R_s}\frac{\sinh r}{r}, \quad \sigma_s(r) = C - \mu c_s(r) + \frac{\mu}{6}\tilde{c}r^2, \tag{11.49}$$

where $C = \frac{\gamma}{R_s} + \mu - \frac{\mu\tilde{c}R_s^2}{6}$, *and* R_s *is the unique solution of the equation*

$$\tanh R_s = \frac{R_s}{1 + \left(\frac{\tilde{c}}{3\tilde{c}}\right)R_s^2}.$$

The following theorem due to Fontelos and Friedman (2003) demonstrates that the radially symmetric solution becomes unstable and develops 'fingers' as the parameter $M(=\mu/\gamma)$ is increased. The corresponding result in 2D was proved by Friedman & Reitich (1999).

Theorem 11.1.C *(Fontelos and Friedman, 2003)*
Let

$$M_\ell(R) = \frac{(\ell-1)\ell(\ell+2)}{2}\frac{1}{R^5 P_0(R)\{P_1(R) - P_\ell(R)\}}, \tag{11.50}$$

where

$$P_n(r) = \frac{I_{n+\frac{3}{2}}(r)}{rI_{n+\frac{1}{2}}(r)}, \tag{11.51}$$

$I_m(R)$ *is the modified Bessel function given by*

$$I_m(r) = \sum_{k=0}^{\infty}\frac{(r/2)^{m+2k}}{k!\Gamma(m+k+1)}. \tag{11.52}$$

For any $\ell \geq 2$, *there exists a stationary solution with free boundary*

$$r = R + \varepsilon Y_{\ell,0}(\theta) + O(\varepsilon^2) \tag{11.53}$$

$$M = M_\ell + \varepsilon M_{\ell,1} + O(\varepsilon^2), \quad M = \mu/\gamma \tag{11.54}$$

for any small $|\varepsilon|$. $Y_{\ell,0}(\theta)$ *is the spherical harmonic of mode* $(\ell,0)$, *namely,*

$$Y_{\ell,0}(\theta) = \sqrt{\frac{2\ell+1}{4\pi}}P_\ell(\cos\theta), \quad P_\ell(x) = \frac{1}{2^\ell\ell!}\frac{d^\ell}{dx^\ell}(x^2-1)^\ell. \tag{11.55}$$

References

Bazaliy, B. and Friedman, A. (2003) 'A free boundary problem for an elliptic-parabolic system: Application to a model of tumor growth', *Communications in Partial Differential Equations* **28**, 517–560.

Chen, X. and Friedman, A. (2003) 'A free boundary problem for an elliptic-hyperbolic system: An application to tumor growth', *SIAM Journal of Mathematical Analysis* **35**, 974–986.

Cui, S. and Friedman, A. (2003) 'A hyperbolic free boundary problem modeling of tumor growth', *Interfaces and Free Boundaries* **5**, 159–182.

Dorie, M. J., Kallman, R. F., and Coyne, M. A. (1986) 'Effect of cytochalasin b, nocodazole and irradiation on migration and internalization of cells and microspheres in tumor cell spheroids', *Experimental Cell Research* **166**, 370–378.

Dorie, M. J., Kallman, R. F., Rapacchietta, D. F., van Antwer, D., and Huang, Y. R. (1982) 'Migration and internalization of cells and polystyrene microspheres in tumor cell spheroids', *Experimental Cell Research* **141**, 201–209.

Fontelos, M. A. and Friedman, A. (2003) 'Symmetry-breaking bifurcations of free boundary problems in three dimensions', *Asymptotic Analysis* **35**, 187–206.

Friedman, A. (2006) 'Cancer models and their mathematical analysis', in *Tutorials in mathematical biosciences III* (A. Friedman, ed.). Springer-Verlag, Berlin. 223–246.

Friedman, A. and Reitich, F. (1999) 'Analysis of a mathematical model for growth of tumors', *Journal of Mathematical Biology* **38**, 262–284.

Hanahan, D. and Weinberg, R. A. (2000) 'The Hallmarks of Cancer', *Cell* **100**, 57–70.

Jiang, Y., Pjesivac-Grbovic, J., Cantrell, C., and Freyer, J. P. (2005) 'A Multiscale Model for Avascular Tumor Growth'. *Biophysical Journal* **89**, 3884–3894.

Pettet, G., Please, C. P., Tindal, M., and McElwain, D. (2001) 'The migration of cells in multicell tumor spheroids', *Bullation of Mathematical Biology* **63**, 231–257.

Ribba, B., Colin, T., and Schnell, S. (2006) 'A multiscale mathematical model of cancer and its use in analyzing irradiation therapies', *Theoretical Biology and Medical Modelling* **3**, 7.

Webb, G. F. (1986) 'Logistic models of structured population growth', *International Journal of Computational Mathematic and Applications* **12A**, 527–529.

Webb, G. F. (1987) 'An operator-theoretic formulation of asynchronous exponential growth', *Transactions of the American Mathematical Society* **303**, 751–763.

Exercises

1. Consider a population of cells in different phases of the cell cycle. Let a denote the age of a cell in the cell cycle ($a = 0$ at birth and $a = A$, where A is the time when the cell divides). The number density of cells with cell-cycle age a at time t is symbolized by $n(a,t)$. If $\mu(a)$ is the death-rate coefficient of these cells, then

$$\frac{\partial n(a,t)}{\partial t} + \frac{\partial n(a,t)}{\partial a} = -\mu(a)n(a,t)$$

for $0 < a < A$, $t > 0$, and $n(0,t) = 2n(A,t)$,

where the last expression accounts for cell doubling. Given $n(a, 0) = g(a)$, compute $n(a, t)$ for all $t > 0$ for the case $\mu(a) = \text{constant}$, and determine the asymptotic behavior of $n(a, t)$ as $t \to \infty$. (For additional models of age-dependent population growth, see the articles of Webb (1986, 1987)).

2. Prove the formula 11.36.
3. Verify eqn 11.49 and prove that there is a unique solution R_s to the equation

$$\tanh R_s = \frac{R_s}{1 + (\tilde{c}/3\bar{c})R_s^2} \quad \text{if } \tilde{c} < \bar{c}.$$

Prove that if $\tilde{c} > \bar{c}$ then there is no solution to the preceding equation. Explain the biological implication of this result for the system 11.41–11.47.

4. A tumor cannot grow beyond a few millimeters without a new supply of nutrients such as oxygen. If such a supply is blocked (by a drug, for instance), then tumor growth is arrested (size becomes stationary) and necrotic cells populate the tumor core (also referred to as 'dead core'). A simple spherical model of tumor with radius $r = R$ and concentric spherical necrotic core of radius ρ is given by

$$-\nabla^2 c + c = 0 \quad \text{in } \rho < r < R$$
$$c(r, t) = c_0 \quad \text{if } 0 < r < \rho$$
$$c = \bar{c} \quad \text{on } r = R$$
$$\int_\rho^R (c - \tilde{c})r^2 dr = 0.$$

Verify that a solution is given by

$$c(r) = \frac{\bar{c}}{r}\sinh(R - \rho) + \rho \cosh(r - \rho) \quad \text{for } \rho < r < R,$$

where R and ρ satisfy the equations

$$\sinh(R - \rho) + \rho \cosh(r - \rho) = \frac{\bar{c}}{c_0}R$$
$$(R - \rho)\cosh(r - \rho) + (R\rho - 1)\sinh(R - \rho) = \frac{\bar{c}}{3c_0}(R^3 - \rho^3) + \frac{\mu}{\sigma_0}\rho^3.$$

Solve these equations numerically for the case $\tilde{c} < c_0$.

Glossary

acetylation. A chemical change that involves the replacement of a hydrogen atom (H) by an acetyl group CH_3CO.

actin. A contractile protein found in muscle cells. Together with myosin, actin provides the mechanism for muscle contraction.

adenoma. Benign tumor arising in glandular epithelium.

adenocarcinoma. A cancer that develops in the glandular lining of an organ such as the lungs.

anaphase. The stage in mitosis in which the chromosomes begin to separate.

angiogenesis. Growth of new blood vessels. Tumor angiogenesis is the growth of blood vessels from surrounding tissue to a solid tumor. This is caused by the release of chemicals by the tumor.

antibody. A type of protein made by certain white blood cells in response to a foreign substance (antigen). Each antibody binds to a specific antigen.

anticodon. A 3-base sequence in a tRNA molecule that base-pairs with its complementary codon in an mRNA molecule.

antigen. Any foreign substance, usually a protein, that stimulates the body's immune system to produce antibodies.

apoptosis. One of the two mechanisms by which cell death occurs (the other being the pathological process of necrosis). Apoptosis is the mechanism responsible for the physiological deletion of cells and appears to be intrinsically programmed.

ADP. Adenosine diphosphate

ATP. Adenosine triphosphate

autocrine signals. Signals that affect only cells of the same cell type as the emitting cell.

biomolecule. Substance that is synthesized by and occurs naturally in living organisms.

carcinoma. A malignant tumor that begins in the lining layer (epithelial cells) of organs. At least 80% of all cancers are carcinomas.

cell senescence. The stage at which a cell has stopped dividing.

cellular automata. Simplified mathematical models of spatial interactions, in which sites or cells on a landscape are assigned a particular state, which then changes stepwise according to specific rules conditioned on the states of neighboring cells.

centriole. An organelle in many animal cells that appears to be involved in the formation of the spindle during mitosis.

centromere. A specialized chromosome region (the constraint 'waist' of the chromosome) to which spindle fibers attach during cell division.

centrosome. A dense body near the nucleus of a cell that contains a pair of centrioles.

charged tRNA. Transfer RNA molecule bound to an amino acid.

chromatids. Each of the two daughter strands of a duplicated chromosome joined at the centromere during dell division.

chromatin. The chromosome as it appears in its condensed state, composed of DNA and associated proteins (mainly histones).

chromosome. A threadlike linear strand of DNA and associated proteins in the nucleus of animal and plant cells that carries the genes and functions in the transmission of hereditary information.

codon. A particular sequence of three nucleotides in mRNA coding for an amino acid.

cytochrome c. A protein present in mitochondrial membranes, it is important in the energy-generation machinery of the cell.

cytokine. Any of many soluble molecules that cells produce to control reactions between other cells.

cytokinesis. The division of the cytoplasm of a cell following division of the nucleus.

cytoplasm. The contents of a cell, outside of the nucleus.

cytoskeleton. A fibrous network made of proteins that contributes to the structure and internal organization of eukaryotic cells.

cytosol. The fluid portion of the cytoplasm, which is the part of the cell outside the nucleus.

DNA polymerase. An enzyme that catalyzes synthesis of a DNA under direction of single-stranded DNA template.

ectopic expression. The expression of a gene in an abnormal place of an organism. This can be caused by a disease, or it can be artificially produced as a way to help determine what the function of that gene is.

embryonic stem cell. Embryonic stem cells (ES cells) are stem cells derived from the inner cell mass of an early-stage embryo known as a blastocyst. Human embryos reach the blastocyst stage 4–5 days past fertilization, at which time they consist of 50–150 cells. ES cells have the potential to become a wide variety of specialized cell types.

endoplasmic reticulum (ER). An extensive network of internal membranes within an eukaryotic cell that is necessary for protein synthesis.

epigenetic. An epigenetic change does not change the sequence of DNA bases but may indirectly influence the expression of the genome.

epithelial cell. A cell that covers a surface of the body such as the skin or the inner lining of organs such as the digestive tract.

eukaryotic cell. A cell containing a nucleus.

exon. A segment of a gene that contains instructions for making a protein. In many genes the exons are separated by 'intervening' segments of DNA, known as introns, which do not code for proteins; these introns are removed by splicing to produce messenger RNA.

fibroblast. Common cell type, found in connective tissue, that secretes an extracellular matrix rich in collagen and other macromolecules. These cells migrate and proliferate readily in wound repair and in tissue culture.

gene operator. A segment of DNA that regulates the activity of the structural genes of the operon that it is linked to, by interacting with a specific repressor or activator. It is a regulatory sequence for shutting a gene down or turning it 'on'.

gene promoter. A region of DNA that is located upstream (towards the 5′ region) of the gene that is needed to be transcribed.

genome. All of the genetic information or hereditary material possessed by an organism.

genomics. The study of an organism's entire genome.

germ cell. Reproductive cells; the egg and sperm cells.

glioma. A cancer of the brain that begins in glial cells (cells that surround and support nerve cells).

growth factor. A substance that influences growth by changing or maintaining the rate that cells divide.

hematopoiesis. The process of formation, development, and differentiation of the cells of whole blood.

histone. A type of protein found in chromosomes; histones attached to DNA resemble 'beads on a string.'

housekeeping genes. Constitutively expressed genes. Housekeeping genes are continuously transcribed at low basal levels.

hypoxia. A condition in which there is a decrease in the oxygen supply to a tissue.

in vitro. In the laboratory (outside the body). The opposite of *in vivo* (in the body).

in vivo. In a living organism, as opposed to *in vitro* (in the laboratory).

inflammation. Redness, swelling, pain, and/or a feeling of heat in an area of the body. This is a protective reaction to injury, disease, or irritation of the tissues.

interactome. The whole set of molecular interactions in cells.

interphase. The portion of the cell cycle where the cell is not dividing; includes G1, S and G2 stages.

intron. 'Intervening sequence,' a stretch of nucleic-acid sequence spliced out from the primary RNA transcript before the RNA is transported to the cytoplasm as a mature mRNA; can refer either to the RNA sequence or the DNA sequence from which the RNA is transcribed.

ion channel. A protein integral to a cell membrane, through which selective ion transport occurs.

kinase. An enzyme that adds phosphate groups to proteins.

kinetochore. The region at which the microtubules of the spindle attach to the centromeres of chromosomes during nuclear division.

leukemia. Cancer that starts in blood-forming tissue such as the bone marrow, and causes large numbers of blood cells to be produced and enter the bloodstream.

ligand. A soluble molecule such as a hormone that binds to a receptor.

lymphocyte. A white blood cell. Present in the blood, lymph and lymphoid tissue.

lymphoma. A tumor of the lymphatic system.

mammal. A warm-blooded animal that has hair on its skin and whose offspring are fed with milk secreted by the female mammary glands.

meiosis. Cell division by which eggs and sperm are produced. Each of these cells receives half the amount of DNA as the parent cell.

metabolite. Any intermediate or product resulting from metabolism.

metabolome. All native metabolites, or small molecules, that are participants in general metabolic reactions.

metaphase. A stage in mitosis or meiosis during which the chromosomes are aligned along the equatorial plane of the cell.

metastasis. In cancer, this is the migration of cancer cells from the original tumor site through the blood and lymph vessels to other tissues.

methylation. The addition of a methyl group ($-CH_3$) to a molecule.

microtubule. Long, cylindrical polymer composed of the protein tubulin. It is one of the three major classes of filaments in the cytoskeleton.

mitochondria. Structures in the cell that generate energy for the body to use. Mitochondria are called the powerhouses of the cell.

mitogen. A substance that induces cell division.

mitosis. The process of division of somatic cells in which each daughter cell receives the same amount of DNA as the parent cell.

mitotic spindle. A network of microtubules formed during mitosis. These microtubules attach to the centromeres of the chromosomes and help draw the chromosomes.

Monte Carlo simulation. A method that estimates possible outcomes from a set of random variables by simulating a process a large number of times and observing the outcomes.

multipotent stem cell. Class of stem cells that can differentiate into more than one tissue type, but not all.

mutation. A change in the number or arrangement of the sequence of DNA.

necrosis. A type of cell death in which cells swell and break open, releasing their contents and can damage neighboring cells and cause inflammation.

nuclear lamina. Dense, fibrillar network composed of intermediate filaments made of lamin that lines the inner surface of the nuclear envelope in animal cells.

nucleoid. The aggregated mass of DNA that makes up the chromosome of prokaryotic cells.

nucleolus. Structure in the nucleus where ribosomal RNA is transcribed and ribosomal subunits are assembled.

nucleotide. The basic unit of DNA or RNA, consisting of one chemical base, a phosphate group, and a sugar molecule.

oocyte. Unfertilized egg cell.

operon. A unit of genetic material that functions in a co-ordinated manner by means of an operator, a promoter, and one or more structural genes that are transcribed together.

oxidative stress. A condition of increased oxidant production in animal cells characterized by the release of free radicals, resulting in cellular degeneration.

phenotype. The observable traits or characteristics of an organism, for example hair color, weight, or the presence or absence of a disease.

phosphatase. An enzyme that removes a phosphate from a nucleic acid or protein.

phospholipids. Any of various phosphorus-containing lipids, such as lecithin, that are composed mainly of fatty acids, a phosphate group, and a simple organic molecule such as glycerol. Phospholipids are the main lipids in cell membranes.

phosphorylation. The chemical addition of a phosphate group to a protein or another compound.

polypeptide. Several amino acids linked together by a peptide bond.

pluripotent stem cell. A stem cell that can form any and all cells and tissues in the body.

prokaryotic cell. A cell having no nuclear membrane and hence no separate nucleus.

prophase. First stage of mitosis during which the chromosomes are condensed but not yet attached to a mitotic spindle.

proteome. The collection of all proteins in the body of an organism. For humans, it is estimated that there are 250 000–300 000 different proteins, of which fewer than half have been catalogued thus far.

retinoblastoma. A malignant tumor that forms on the retina. Retinoblastoma most often affects children under the age of 5.

ribosomes. Small cellular components composed of specialized ribosomal RNA and protein; site of protein synthesis.

RNA polymerase. An enzyme that synthesizes RNA, usually from a DNA template.

sarcoma. Malignant tumor arising in the bone, cartilage, fibrous tissue or muscle.

somatic cell. Any type of cell other than the reproductive cells (egg or sperm).

stem cell. An undifferentiated cell that possesses the ability to divide for indefinite periods in culture and may give rise to highly specialized cells of each tissue type.

stroma. The supporting framework of an organ, typically consisting of connective tissue.

structural gene. A gene that controls the production of a specific protein or peptide.

telomeres. Repeated DNA sequences found at the ends of chromosomes; telomeres shorten each time a cell divides.

transcription factor. A DNA-binding protein that regulates expression of a gene.

transcriptome. The full complement of activated genes, mRNAs, or transcripts in a particular tissue at a particular time.

transfer RNA (tRNA). A class of RNA having structures with triplet nucleotide sequences that are complementary to the triplet nucleotide coding sequences of mRNA. They carry amino acids into ribosomes for protein production.

tubulin. Most abundant protein in microtubules.

ubiquitin. A protein found in all eukaryotic cells that becomes covalently attached to certain residues of other proteins. The attachment of a chain of ubiquitins tags a protein for intracellular proteolytic destruction.

zygote. The cell resulting from the union of sperm and egg.

Index